THE LIMBO FILES

Other Books by David Langford

Fiction
An Account of a Meeting with Denizens of Another World, 1871
Earthdoom! (with John Grant)
* *Different Kinds of Darkness: Short Stories*
The Dragonhiker's Guide to Battlefield Covenant at Dune's Edge: Odyssey Two
* *Guts* (with John Grant)
* *He Do the Time Police in Different Voices: SF Parody and Pastiche*
Irrational Numbers
* *The Leaky Establishment*
A Novacon Garland
* *The Space Eater*

Nonfiction
The Apricot Files: The "Disinformation" Columns
* *The Complete Critical Assembly*
A Cosmic Cornucopia (with Josh Kirby)
Critical Assembly
Critical Assembly II
The End of Harry Potter?
Facts and Fallacies: A Book of Definitive Mistakes and Misguided Predictions (with Chris Morgan)
Micromania: The Whole Truth About Home Computers (with Charles Platt)
The Necronomicon (with George Hay, Robert Turner and Colin Wilson)
Pieces of Langford
Platen Stories
The Science in Science Fiction (with Peter Nicholls and Brian Stableford)
* *The Sex Column and other misprints*
The Silence of the Langford
The Third Millennium: A History of the World AD 2000-3000 (with Brian Stableford)
The TransAtlantic Hearing Aid
The Unseen University Challenge: Terry Pratchett's Discworld Quizbook
* *Up Through an Empty House of Stars: Reviews and Essays 1980-2002*
War in 2080: The Future of Military Technology
The Wyrdest Link: The Second Discworld Quizbook

As Editor
The Encyclopedia of Fantasy (with John Clute, John Grant, and others)
* *Maps: The Uncollected John Sladek*
Wrath of the Fanglord

* In Cosmos Books

THE LIMBO FILES

Writing, Freelancing, and the Amstrad PCW

David Langford

Cosmos Books • 2009
An imprint of **Wildside Press**

THE LIMBO FILES

Published by

Cosmos Books. an imprint of Wildside Press
www.cosmos-books.com
www.wildsidepress.com

Copyright © 2009 by David Langford. All rights reserved.
Original appearances copyright © 1986-2002 by David Langford.
Cover design copyright © Juha Lindroos, 2009.
This Cosmos Books edition published 2009.

The right of David Langford to be identified as the author of this work has been asserted by the Author in accordance with the British Copyright, Designs and Patents Act 1988.

No portion of this book may be reproduced by any means, mechanical, electronic or otherwise, without first obtaining the permission of the copyright holder.

For more information, contact Wildside Press.

ISBN: 978-0-8095-7324-0

Dedication

Once again for Chris Priest, my long-suffering partner in the doomed software company Ansible Information.

Contents

Introduction	9
In the Beginning	11
Front Page Treatment	12
The Fading of the Keys	15
Dinosaurs and Dastards	17
C.E.T.I.	19
Through the Barrier	20
Why Software Isn't Cheap	22
How to Annoy Software Customers	24
The Joy of Mailbags	26
Copyrights & Wrongs	28
Arguments Against	30
Dirty Words	32
Escape Plans	33
Speaking in Tongues	35
Reading for Profit	37
The Paperless Future	39
Computer Plots to Avoid	41
The Book of All Knowledge	43
Mythical but True	45
The Horror in the VAT	47
Runes of Power	48
The Leper's Squint	50
Another Useful Review	52
Best Foot Forward	54
Bits and Pieces	56
Freelance Finances	57
Misleading Cases	59
Public Accolades	61
Down the Line	63
Writing More Gooder	65
100 Years Ago	66
Nebulous Statistics	68

The Long Goodbye	70
Fizz! Buzz!	72
Paperless Publishing	73
Little Dots and Squiggles	75
Occupational Diseases	77
Behind Closed Doors	79
Contracting Universe	81
Electromagnetic Etiquette	82
Post-1989 Miscellany	84
Swings & Roundabouts	86
Stalking the Wild Editor	88
Software Deathwish	90
UFO Follies	91
Amstrad PPC Blues	93
Dangerous Corners	94
Aftermath of Glory	96
Dictionary of Quotations	98
Getting Together	99
Half Century	101
Legal Frictions	103
Hack's Quest	105
Tweedledum and Tweedledos	108
Index, Peculiar	110
Laser Duel	112
Bye-Bye BT	114
The Knowledge Trap	116
Desperate Times	117
Going Commercial	119
The Padded Cell	122
Losing Your Chains	124
Old Joyce's Almanack	126
Ticking the Boxes	128
False Prophets	130
Enquire Within Upon Everything	132
Eureka! The Wheel!	134
How Not to Write	136
Future Shock	138
Golden Phrases	140

Those Crazy Ideas	142
News on the March	144
Caught in the Net	146
In the Beginning	147
Netted Again	149
Dated Information	150
Centenary Blues	152
Party Time	153
Mysteries in the Mail	155
Urban Folklore	156
Gadgets of Yesteryear	157
My Fantasy Life	159
Life Forms	160
At Another Crossroads	161
Unspeakable Researches	163
Critical Mass	164
Baron Munchausen Remembers	165
The Customer is Always Right	167
Thog's Masterclass	168
Looking Backwards	170
Change and Decay	171
Take Another Look	173
Money, Money, Money	174
Under the Bonnet	176
Pig in a Poke	178
Me, *Ansible*, and Thog	180
I Didn't Write That!	181
Running Down	183
Index	185

Introduction

As promised by the subtitle, this is a collection of ramblings about writing (of which I've done a little bit), freelancing (in which sordid calling I've somehow survived since 1980) and the Amstrad PCW word processor – which had to be mentioned from time to time since it was the official subject of the magazines where all these columns appeared.

For someone who – at least according to me – is desperately shy about self-promotion, I seem to have had considerable luck at worming my way into magazines as a regular columnist. I dimly remember coming across a small and tatty stand for Future Publishing's *8000 Plus* at a London computer show during the summer of 1986, and, having presumably had a few beers at lunchtime, assuring the editor that his future would be dust and ashes without the soothing oasis of a Langford column. When the first issue appeared in October, there I was – not to be completely dislodged (despite one hiatus from July 1993 to May 1994) until the ailing magazine was at last axed at the end of 1996.

By then it was called *PCW Plus*, since *8000 Plus* had been intended to echo the code numbers of Amstrad's two dedicated word processors, the PCW (Personal Computer Word-processor) 8256 and the PCW 8512 – only for Amstrad to complicate matters by later adding the PCW 9512 and others. The last three digits were measures of built-in memory, and modern readers may gasp at the realization that this meant 256 and 512 kilobytes, not megabytes. Amstrad's allegedly lovable Cockney boss Alan Sugar had wanted the cheapest possible system, so all these early PCW machines had elderly 8-bit processors, unfashionable 3" disk drives, and the utterly venerable CP/M operating system (Control Program for Microcomputers). Only dedicated nerds like myself ventured into the deep user-unfriendliness of CP/M, since the few simple things a writer needs from the computer's operating system – copying documents, formatting disks and the like – were all built into the custom word processor, LocoScript. This was written for Amstrad by Locomotive Software, who later rebranded as LocoScript Software.

It's easy to forget how revolutionary this bare-bones word processor package seemed, and how madly popular it became in Britain: affordable, (fairly) simple, and with no extras to buy. The computer hardware and floppy drive or drives were built into the green monochrome monitor, the keyboard and supplied printer couldn't be plugged in wrongly, and although the distinctive matrix-print output of LocoScript 1 was mildly horrible it was very much better than no word processor at all. There was a time when every science fiction fanzine in the country seemed to be produced on a PCW. Ah, nostalgia.

Because the PCWs were marketed as word processors rather than computers, I was allowed to write for writers about writing in what was nominally a computer magazine, with only occasional lapses into technicality. It was a lot of fun. Even after *PCW Plus* shuffled off its mortal coil, I was persuaded to make a comeback in the small-press magazine *PCW Today* (see page 170), where the column straggled on for a few years despite the undeniable fact that PCW-series computers were dead, dead, dead.

Some of the distilled wisdom here is of course a mite outdated -- for example, the June 1987 'Copyrights & Wrongs' column on page 28 was written before the USA finally signed up to the Berne Convention in 1989 – but I like to think that a good deal of the advice and comment on writing and freelancing is timeless. The jokes certainly are. A very few update notes have been added in italics and square brackets *[like this]*.

Finding a title for this collection required much head-scratching. I started with the deeply boring *The Amstrad PCW Files*, as a companion to my much more technical and user-hostile collection of Apricot computer columns, *The Apricot Files* (Ansible Information, 2007). Then I tried to brighten it up as *My Secret Life with Alan Sugar* and assorted variations on that theme. Finally I begged advice from the great brains of the Usenet newsgroup rec.arts.sf.composition, where Graham Woodland promptly suggested *The Limbo Files* – PCW Limbo being the retirement home for deleted LocoScript files, as mercilessly explained on page 86. Thanks!

Rather than leave the rest of this page blank, I here insert the 'Sucrose Patter Song' from a reprehensible column of computer-updated Gilbert and Sullivan lyrics which I contrived to publish in *New Computer Express* in 1989. For an explanation of the mysterious reference to tactical nuclear weapons, turn to page 33.

> *If you ask me how to rise in this hardware enterprise and become a mega-star,*
> *I say, fill the papers' pages with some sentiments outrageous, never mind just what they are.*
> *So a scheme that's hardly practical for selling weapons (tactical) means column space for free –*
> *Then in headlines that are shockers you may castigate the 'knockers', and it's all publicitee!*
> *– And everyone will say,*
> *As you walk your flagrant way,*
> *'If that young man gets quoted even on the BBC,*
> *Why, what a most particularly famous man that bearded man must be.'*
> *Then does your product need improving? Still you keep the boxes moving and don't ever tell a soul:*
> *But ensure each media mention is designed to shift attention from your quality control.*
> *Though the magazines may grumble, you will never take a tumble while the punters all admire*
> *Your delightfully outspoken words of confidence unbroken – in a cockney accent, squire.*
> *– And everyone will say,*
> *As you walk your rough-hewn way,*
> *'If that young man needs elocution lessons just like me,*
> *Why, what a most particularly unspoilt man that bearded man must be!'*

So here at last is the collected edition of 'Langford's Printout', an editor-imposed column title which later became the austerely confident 'Langford'. It is published solely for corrupt personal gain. To quote, not for the first time, a favourite line from *The Treasure Seekers* by E. Nesbit: 'If what we have written brings happiness to any sad heart we shall not have laboured in vain. But we want the money too.'

David Langford, January 2009
http://ansible.co.uk/

In the Beginning

Well, it's the old, old story: like 5,271,009 people in the Home Counties alone, I bought a PCW 8256. My motives were half noble and pure (I'm a professional writer), half hopelessly corrupt: I also prey on unfortunate punters by flogging them software. Your editor thinks my jottings must therefore be deeply interesting to you all. Little does he know.

The other day I was slumming in an IBM PC magazine, where I found the fascinating news that Amstrad PCW systems have met with a 'lukewarm response'. To translate this you need to know the subtle linguistic codes used by IBM enthusiasts. If you buy an IBM PC, that's a ringing declaration of total commitment. If you buy something else, it's a lukewarm response.

Back in reality, a mate of mine at Victor Gollancz, the publishers, reports that although the number of truly terrible magazines he receives hasn't altered *that* much, there's been a distinct change of appearance. Time was when all too many first novels were handwritten in blue crayon in exercise books. Now fully 25% of them are LocoScripted in that unmistakable Joyce typeface. *[Newer readers may need to be reminded that before the coming of the PCW 9512 and of LocoScript 2, all PCW print-outs looked exactly the same. And that Alan Sugar's secretary Joyce provided the machine's working name.]*

Once upon a time it was a treasured piece of publishing folklore that a beautifully printed submission with multiple type-styles and straight margins at both left and right, exquisitely bound in tooled goatskin with rich puce endpapers ... was never any good. But since the coming of cheapo word-processing, a 'perfect' typescript is no longer evidence of pointless obsession. The twiddly bits have ceased to be a problem; reprinting a corrected page is a doddle. There are, however, still points to watch.

Don't overdo it. Avoid hordes of different typestyles. If a long passage has to go in italics, it's probably OK to print it that way (as a reminder to typesetters some professionals also write 'italics' in the margin). However, using LocoScript italics for odd emphasized words may be a mistake, depending on the publisher. Copy-editors in traditional firms will hate you because they'll need to underline all those words for the typesetters – better that you save them the trouble. Copy-editors in modern dynamic firms may well be able to transfer your work straight from disk, and will prefer the italics to be italics from the start. Ask first.

In any case, avoid justified right-hand margins; most editors feel these look subtly naff in typescripts.

Don't forget to use the handy 'header' facility to number every page. At top right is a good place. Include your surname and a short form of the book's title, too. Almost no extra work is involved, and you'll have insured against the day when some drunken editor accidentally drops and shuffles the pages of your masterpiece with several other identically LocoScripted (but of course deeply inferior) submissions.

Don't muck around with any gadget, device or process which glues, screws, wires, welds or staples hundreds of MS pages into a single 'book'. Publishers hate the massive objects: they're awkward to handle and usually get ripped apart for convenience (as will in any case happen at the typesetting stage, unless you can persuade them to set directly from your disks). Loose A4 pages in a box are quite acceptable; it's permissible to use staples of paperclips at top left to fasten together individual chapters or sections.

Don't, after spending all this loot on your word processor, skimp on the paper. If you try to save paper by printing at 17 characters per inch, or ribbon by sticking to 'draft' quality, your editor will enclose a hefty bill from his optician along with the rejection slip. (High quality print with 10 or at most 12 characters per inch, please.) If you don't double-space your print-out with +LS2 it probably won't even be read. Ditto if you're such a miser as to leave only a quarter-inch margin all round the edge: allow lots of space for editors to make their cryptic marginal notes, such as 'By George, what a work of genius!'

And finally, use paper with a certain amount of moral fibre. Gossamer-thin, electronically-tested stuff which wilts and slithers all over the editorial desk may not ensure rejection, but can tip the balance when someone is debating whether to finish your MS or go to the pub....

Oddity of the Month

Try this LocoScript experiment: hold down the Shift key and type 'OK'. No, *really fast*, so the O key hasn't quite risen before the K goes down. A spurious Return magically appears between the two letters. As a rapid typist, I wondered for a while whether LocoScript had a built-in style checker and was conditioning me to type the preferred 'Okay' or perhaps 'Yea, verily, forsooth.' However, it happens in CP/M and other programs too: there is apparently a bug in the PCW's keyboard scanning. Not quite as bad as the time when I was trying to run a Prestel SF database from a Commodore 64 and couldn't speed-type the letters SF – they persistently came out as SXF.

8000 Plus 1, October 1986

Front Page Treatment

Wearing my other hat as an SF reviewer, I waste my days reading lots of allegedly imaginative new books featuring computers. The megacomputers of SF have several irritating points in common:

They all *work*, and never beep at you for failing to slot in the proper 'start of day' disk with the hidden LocoScript program files. (By the way, you can save 4k of space on your LocoScript start-of-day disk by displaying – with **f8** 'Show Hidden' – and then erasing the file MAIL232.COM, a communications program which is no earthly use until you buy an interface box. Keep the master copy....) They work fast, depriving you of the fun of racing trained slugs along a measured 800 metres while LocoScript chugs to the end of a document. Most SF computers talk or take dictation, or can even be plugged directly into the brain so that your thoughts flash straight to the screen ... either way there's no trouble with coarse, mundane keyboards.

Back in reality we have the PCW, which at least doesn't devour its users' brains (though my wife isn't so sure), take over the universe (even if they currently seem to be outbreeding mere humans) or precipitate nuclear holocaust (I know both CND and Greenpeace people with Amstrads, but they're oddly rare in both the

Pentagon and the Kremlin).

The worst SF threat that the beasties pose is the Creeping White Peril, subtitled the Paper That Ate Manhattan. You buy the gadgetry with vague thoughts of that high-tech catchphrase 'the paperless office'. A week later you're entirely surrounded by crumpled manuals and early drafts ... all around the monitor are stuck Post-It Note reminders like 'Press Extra 4 for ¢ sign' (Loco 2: Shift Alt **4**) or 'to alter page numbers don't press **f6** for Pages but **f7**, Enter, long pause, **f7**, **f8**' (Loco 2: don't press **f5** but **f1**, Enter, **f5**) ... and the floor is a sea of discarded sheets because brilliant on-screen sentences can lose their shine when printed out, forcing you to redo page after page.

You do need to cultivate an eye for the screen. Publishers' readers train their eye in a different way: they're the poor sods who first sift through submitted typescripts in search of some glimmer of literary virtue, and the task is complicated by the psychological fact that in mere typescript nothing looks as good as Real Print. On the screen we have the reverse problem, maybe because we're conditioned by all that SF about infallible super-machines: those neat green words look so convincing that your eye skips clean over the typpong errors and warped; punctuation. Learn to be distrustful.

Some good news on the monitor front is that despite horror stories of deadly screen radiations which turn you into a low-budget special effect from *Doctor Who*, current researches indicate that computer monitors are somewhat less dangerously radioactive than houses or people. Must tell this to the software customer who's convinced his screen emits cosmic rays, making him nauseated whenever he sits down to work. I used to get the same sensation from mere pencil and paper – the central problem wasn't so much the dread A4-radiation and HB-particles as working for the civil service. A letter of resignation produced a complete cure. Glory, glory.

There are two valid worries about the PCW monitor. One is very general: if you fret about evil effects of working with the green screen, the mere stress of worrying can itself be bad for you – so don't. (Take my elixir and you'll live forever, provided you never once think of the word 'hippopotamus'.) Eyestrain is the more specific worry. You should try to stop staring at the screen for at least ten minutes of each hour, which isn't difficult if you regularly save your files to disk so that LocoScript can do its famous simulation of an interglacial period.

Researching kindness to my own eyes, I find that my PCW display looks most readable with the brightness at absolute minimum, and I wouldn't mind turning it down further: maybe one day I'll invalidate my guarantee by soldering in a resistor. If anyone out there beats me to this bit of illicit tinkering, do give me a call from the intensive care ward and say how you got on.

Meanwhile, that horrid coughing from offstage is your editor reminding me to mention writing. Last month I revealed some open secrets about LocoScripting submissions to publishers: more follows, but do please remember one point. Every new writer has to confront the embarrassing fact that no matter how many tips on presentation you amass, you will eventually have to display some actual talent.

The virtue of word processing is to mechanize the boring bits, so it's worth setting up a standard front page for your typescripts. Because the most tedious things to type are those known by heart, like your own address, it's horrifyingly easy to me off-putting mistakes right there on the covering sheet. Avoid existential

Angst by using the TEMPLATE.STD facility.

You can stick lots of boilerplate material into this LocoScript 'template' document ... everything, in fact, that you want to have appear in each new document. (Different LocoScript disks and document groups can use different TEMPLATE.STD files or none at all.) A header, for example, a base layout with double-spaced lines (+LS2) and a starting page number. I usually specify no header at all for the first page (the cover sheet) and the magical word 'Langford' plus a page number as a right-justified header for following pages. For the cover sheet I set the page number at zero: this isn't really part of the MS, but carries information for the editor (it'll be ripped off before typesetting).

If you specify your cover sheet and part of page one in TEMPLATE.STD for your 'stories and articles' group, there's no need to lay it out each time: it'll be popped automatically into each brand-new document. A rough suggestion based on my own slovenly habits is to start with ten or so Returns to help centre the text vertically on the cover sheet (you can tinker with this later). Then enter something like the following, centred or spaced out as you see fit:

```
                    TITLE
            Author's Name (i.e. yours)
                x000 words approx
              Not previously published
                 Address line 1
               Address line 2 (etc)
         [NEW PAGE AND SEVERAL RETURNS]
                 TITLE again
                Author's Name
```

... and after a blank line or so, the text begins. As you start a new document you can add the appropriate title; when you've finished and counted the words, stick the figure into its slot on the cover sheet.

Explanations and comments follow. Word counts are vital for short pieces; round them to the nearest 10 or 50 words (100 or 500 for full books), since 'exactly 5751 words' is often regarded with suspicion as indicating an obsessive amateur. If you have a literary agent, his/her address can follow or replace your own. The point of the blank lines before the title on page one is that editors like to have this space to scribble typesetters' directions – how the title is to appear, etc. Some people repeat their address on page one, but the editor will only have to cross it out come typesetting time.

Your mission, should you choose to accept it, is to get your text into print and to have it bear some resemblance to what you actually wrote. Reduce the work editors have to do ... once they get started on a script that needs extensive marking-up and correction, they find it fatally easy to carry on and fiddle around 'improving' your text. And watch out for ambiguities. For example, liberal use of soft hyphens can tidy the right-hand margin – but your editor will hit you in the mouth, having stayed up all night deciding whether each hyphen is meant to be there or should be closed up when the text is rearranged.

What about dashes? One isolated hyphen can get accidentally joined to an adjacent word and give the illusion that hyphenation was intended. It's wiser to type a dash as two consecutive hyphens. (Some writers prefer three.) But during relaying, two 'ordinary' hyphens can break apart at a line-end and look silly ... so use 'hard' hyphens, which stick together, although LocoScript makes them a bore to type. You might think it worth making a pair of hard hyphens into a 'standard phrase' to be entered with, say, the Paste key plus **D**-for-dash.

Questions? Violent disagreements? Letter bombs? Send them to our editor, who is delighted to forward such things after searching the envelopes for money.

8000 Plus 2, November 1986

The Fading of the Keys

One pleasant surprise about the PCW on its first release was the keyboard. Thanks to Clive Sinclair, we all thought that typing on a low-cost machine had to feel like fingering a plastic packet of salami (the ZX80 and ZX81) or squashing small unfortunate sea anemones (the Sinclair Spectrum). Though it's a bit rattly, the 8256/8512 keyboard has a better touch than certain far more expensive computers – even when you go upmarket, typing can still feel like pushing down dead piano keys (the Apricot F series) or clicking tiny retractable ball-pens (all too many IBM compatibles).

The big question staring us in the eyeball today is this: since the keyboard takes more of a bashing from your rough, brutal hands than any other part of the PCW, just how long can you expect it to last? I consulted an early user who has thumped his 8256 a lot more than most: John Grant reckons he's hammered out half a million words of finished text in the first year, including a whopping encyclopaedia of Walt Disney characters. If I mention that he has also collaborated with me on the very wonderful 'ultimate disaster novel' *Earthdoom!* (out next year from Grafton: keep your eye on the Booker Prize list), your editor will get upset at this naked self-promotion and cut me off in mid

Well, this is what famous pseudonym Grant reported. As he came up to the half-million-word mark, two alarming things could be seen happening to his PCW keyboard. The letters on the keys were visibly wearing off (sighs of relief and 'Is that all?' from smug touch-typists), while the keys themselves had a nasty tendency to clog and stick. Imagine all the horrible gunge and dandruff and things from under your fingernails that must collect there over a period of time....

So our hero went round to his dealers, Lasky's, and asked their sage advice. 'No trouble,' said the resourceful Lasky's man. 'The solution's dead simple. You just have to buy a complete new PCW8256 system, that's all.'

This being a family magazine, I am not going to record what Mr Grant replied.

When the apoplexy had died down a bit, he got in touch with compassionate, caring Amstrad themselves. 'Push off,' they told him soothingly. 'We don't deal with the (shudder) general public.' Eventually, worn down by threats and tireless whimpering, they parted with the secret inside information. What you need to do is go to a Real Computer Shop, the sort that doesn't also sell hi-fi systems and cameras, and grovel on your bended knees. (Sometimes this doesn't work and you have to shout as well.) Real Computer Shops are the only people who can order spares from Amstrad ... at least, until some wicked independent manufacturer breaks the unwritten rules by producing its own version for direct sale to scum like you and me and John Grant.

How did it work out? Brace yourself: the price quoted for a new keyboard was a chilly £100. Some time later (i.e. too late) our hero found he could also have the old keyboard repaired, serviced, and generally souped up for £50. After thinking about it, increasing his mortgage, etc., he decided to opt for this as well. It might sound like extravagance, but a professional author can't be caught without a keyboard, any more than a proofreader could get along without the obligatory dark glasses and white stick.

After relating this awesome chronicle, 'John Grant' added that he was investigating a company called Saga which reportedly sells stick-on letters to replace those worn off Spectrum keys by unwholesome practices. 'Seems they're OK on the Amstrad, except of course you don't get the special words for the function keys. I wouldn't be surprised if the company diversified into Amstrad stickers.....'

There are a couple of ways of looking at this. One is that it gives you something to aim for: only when your keyboard fades and disintegrates will you know that you've equalled the output of a real (if somewhat workaholic) author! Alternatively, you can bask in smugness at the thought that at least you got your first keyboard thrown in as part of the package. Incredibly enough, IBM buyers have in the past discovered too late that the keyboard – as well as the monitor and the operating system – was an expensive 'extra'. There was a standing joke about the IBM staff canteen where lunch cost a mere £1.50, plus a pound for the knife, and another for the fork, and another for the plate....

The thing which really worries the more pessimistic PCW users isn't the keyboard (you can always type very lightly and invest in a few sets of the above-mentioned stickers) but the CF2 disks. The fact that they cost three times as much as the 3½" disks for my other computer (and have only half the capacity, unless your PCW has a Drive B) can be lived with. But it seems to be accepted that 3" disks are kept in production almost solely by the Amstrad market. Will Alan Sugar sandbag his existing users – as he's done before, in other ways for other computers – by abandoning 3" drives altogether? Will the disks then become unobtainable? Is the Pope a Scientologist? Rather than make up science-fictional answers to all those queries, I'll merely suggest that it mightn't be a bad idea to stock your fallout shelter with a few spare boxes of CF2s from some discount merchant.

Of course, you may say, you can always use disks again and again. I'm terrifically nervous of doing so: after mere years of word-processing, I know all too well that almost any document may be needed again. For example, a while ago I wrote a short story, titivated it, put in lots of rude bits, cut some of them out again, printed the whole thing, and sold it. I had a clean master copy of the text (photocopied from the print-out – quicker, and cheaper in the long term, than wearing out the PCW printer and ribbon by doing two), plus spare copies of the magazine where it reappeared. A clear case for re-use of disk space! Then a Swedish magazine asked to reprint the story, but there was a snag – could I edit it down by 20% to fit their space restrictions? Which meant keying it in again....

(Also: master copies get spoilt and need reprinting; and from time to time one has second thoughts even about one's own golden prose; and the magazine version is never quite what one wrote; and I've even incorporated short stories into novels. In each case, keeping a copy on disk saves a lot of work.)

In short – according to me, wiping and re-using disks is false economy for any serious writer. Stop the nasty practice at once, or your participles will fall off! There is *no truth* in the rumour that I've just bought a controlling interest in a

major CF2 disk distribution company.
Keyboard Scans Revisited
This column's only devoted fan Andy Lusis sends congratulations on the 'major, ground-breaking report' two issues back – concerning funny effects when you hold down Shift and O and then type K. 'Extensive research on my part has revealed that pressing K before O does the same ... other effects can be produced by pressing CD/DC and KL/LK with Shift in a similar manner. only time will tell what other marvels have yet to be discovered.' For example, when I quickly typed the word Shift just then, it came out as SHIF>T ... the whole upper-case keyboard is riddled with these traps. Try AZ or ZA, which at first gave me a moment of sheer panic – had the PCW hung up? One consolation is that these glitches happen only with Shift held down – not with Shift Lock set.

The New Mathematics
Though I don't usually dabble in BASIC, I couldn't help noticing a lovely revelation in Another (Official) Magazine, which asks you to 'consider the following very simple program: 20 FRED = 30 ... When RUN, all this program does is to make the variable FRED equal to 37.' Clearly BASIC programming is more subtle and complex than I thought.

8000 Plus 3, December 1986

Dinosaurs and Dastards

History has a terrible grip on computer users. These little plastic boxes may be the latest in electronic niftiness, but we're still all stuck with the inefficient QWERTY keyboard ... originally designed to slow people down and avoid stripping the cogs of creaky Victorian typing engines.

There are word-processing equivalents of the QWERTY heritage, the most notorious being the dread WordStar. When computer buffs sneer at LocoScript, they generally drop a smug word about 'real, *industry-standard* word processors like WordStar'. It's a lumbering ten-ton crane of a program, haphazardly designed and hailing from the days before arrow keys ... thus the long-term WordStar fanatic has a deep-seated belief that its traditional Alt-E, Alt-S, Alt-D and Alt-X are good, memorable key combinations for – respectively – Up, Left, Right and Down.

Want to stop editing in WordStar? All you have to do is remember that the 'save' and 'exit' functions are memorably located on the 'Block Moves' menu, which is memorably reached by typing Alt-K. I don't recall just now whether K stands for Save, Exit or Block, but it sure is memorable....

Arthur Naiman, who wrote the best book about this program (*Introduction to WordStar*, published by Sybex), and made sure his contract with the publisher specified that he *needn't* use WordStar to write the book, said this: 'WordStar is one of the most poorly designed word processing programs ever written – a huge, elaborate farrago of kludgy patches, sort of like a Rube Goldberg machine gone berserk. All kinds of basic functions require disk access, thereby making the program fantastically slow....'

('Rube Goldberg' is American for 'Heath Robinson'.)

I've tried to use WordStar and I agree. Yet it lingers, because it's a 'classic', available everywhere, and millions of people have learned its weird ways (new word processors are occasionally slagged off by computer-press hacks for failing to support the 'universally accepted' cursor controls Alt-E, Alt-S ...). Just like QWERTY.

With 'industry standard' opposition like this, don't be ashamed of LocoScript ... even if tempted to take the Locomotive Software boys aside and persuade them with a rubber hose to answer the Ultimate Question of Life, the Universe and Everything. Which is: 'The advantage of a self-loading program which doesn't run under CP/M is that by use of direct hardware and firmware access it can theoretically run with blazing, mindboggling speed. So why...?'

Largely because LocoScript's huge character set requires graphics control of each dot on the screen ... *lots* more time-consuming than just using the 256 off-the-peg CP/M characters. Thus much of its slothfulness comes from the provision of things like Cyrillic and Greek which few people use.

LocoScript could itself become a QWERTY-style dinosaur, because everyone traditionally swears by their first word processor. I have a dark suspicion that IBM versions of LocoScript are being bolted together for those who plan to swap machines and would prefer a familiar program to a more powerful one. *[I was right, too.]*

My own first 'real' word processor (not counting cassette-loaded Scripsit on a Tandy which wouldn't display lower-case, and EasyScript on a CBM64 with the famously naff 40-column screen) was SuperWriter. This is an efficient though slightly outdated program; out of nostalgic fondness I investigated and reviewed the PCW version released by Sorcim.

Something of my disappointment came across in the ensuing article for Another (Official) Magazine ... but not a lot, since the rudest bits were ruthlessly subjected to the CUT key. 'Consummate stupidity ... naffness ... clowns ... wally practices' ... these were among the tactful reproofs which didn't make it into print.

This version of SuperWriter (since withdrawn, and no wonder) is a classic example of an OK program being made almost useless by lousy configuration for the PCW. Instead of adjusting it to the attractive 32x90 screen size, the makers require you to find and run the SET24X80 program before using SuperWriter. The 'status line' information is supposed to be distinguished by reverse video display, but isn't. Likewise the print controls: instead of (+UL) and (-UL), SuperWriter uses a lower-case **u**, distinguished from a normal text **u** by special highlighting ... which got left out of the PCW version! You literally can't tell print controls from real letters unless able to 'patch' the SuperWriter program and substitute distinctive symbols (that's what I did).

SuperWriter appears to have been adapted for the PCW by doing exactly two things: copying a generic CP/M version to 3" disk, and sticking a matrix-printed label on the front of the standard IBM manual to explain that this is the Amstrad version.

It's a pity: I still rather like SuperWriter. Unfortunately the above soul-chilling tale is nothing special in the sleazy world of software. Magazine letter columns are full of wailing about programs which require the 24x80 screen but don't bother to send the two character codes (ESC x) which set up this format; or manuals larded with references to appendices which have been ripped out because 'they don't apply to the Amstrad'.

Next issue, the terrors of communications links. Armed with rod, gun and soldering iron, will our hero achieve brilliant success or second-degree burns? Aha.

8000 Plus 4, January 1987

C.E.T.I.

As every red-blooded SF or space fan knows, CETI stands for Contact with Extraterrestrial Intelligence. It's a good phrase for the awesome concept of using your PCW to signal across the illimitable void of incompatibility – to communicate via feeble pulses of electrons with such hostile entities as Intergalactic Bug-eyed Monsters (another acronym explained) – to boldly go ...

Sorry, the SF writer in me gets a bit uncontrollable at this phase of the moon.

As threatened, I've been looking into ways of squirting text between the PCW and other computers. Suppose you've written a book in LocoScript and your publisher asks for an ASCII file on an IBM 5¼" disk? Or suppose you could (like me) save retyping stuff if you could transfer it directly from 3½" to 3" disks? Assuming you own or can borrow a computer with the 'alien' disk type, this is what you need:

First, your PCW requires the add-on serial interface which lets it talk to the outer world. Amstrad's own isn't too expensive, and can be installed in a few seconds without voiding your guarantee or hauling out the arc-welding gear: it fits on that wafer of circuit board sticking out of the back of the PCW. Two screws hold it there. Technical requirements: one Phillips screwdriver and the ability to rotate same in a clockwise direction.

The interface box announces its presence when you load CP/M, and doesn't otherwise get in your way.

Next, you must physically connect your computer to whatever else is waiting out there. The cables can be quite pricy, and I needed two (we'll come to that), so I decided to go stark mad, buy the bits and solder up my own. This proved much easier than expected, enabling me to save vast sums of beer money. If you can face the thought of using a soldering iron, here's the recipe.

Take a few metres of 3- or 4-core screened cable (only 3 cores will be used) and two of those arcane objects known to the Illuminati as '25-way D connectors', Consult your local electronics shop, or the ubiquitous Maplin catalogue. The PCW end of the cable needs a socket-type ('female') connector; at the far end, IBM-compatibles like the Amstrad PCW 1512 demand the identical connector, while other computers like Apricots may want the plug-type ('male') version. This is why I wanted two cables.

The actual wiring isn't too agonizing, as the pin numbers appear in tiny bas-relief on the connectors. Pin 2 at the PCW end should go via the cable to Pin 3 at the other end. Likewise, Pin 3 to Pin 2. The third cable wire links Pin 7 at each end. Finally wire together pins 4, 5, 6 and 8 at the PCW end, and then do the same at the far end. 'Crossing' the links between pins 2 and 3 produces a cable suitable for talking to other computers rather than mere 'peripherals' like printers.

After connecting this cable between the Amstrad add-on box and the 'serial port' of the alien computer, there's the question of software to move stuff to and fro. No problem with the PCW itself: the famous MAIL232 program is lurking on your master disks and can be commanded by loading CP/M, putting in the

LocoScript disk and typing MAIL232. Further details are in the booklet that comes with the interface.

My first experiment was with an Apricot and its ASYNC communications program, which like MAIL232 was thrown in free. Nothing much seemed to happen in the first few trials, until I tried a lower 'baud rate'. This is a measure of how fast the information is moving down the wire, in bits per second. MAIL232's normal setting is 9600 baud, moving text at 1200 characters per second (one character equals one byte equals eight bits). Perhaps my rotten little wire couldn't cope with the traffic, perhaps the computers just weren't quite compatible at that speed. The problem vanished when I braked to 1200 baud, and text files starting moving across with gratifying ease.

(Many computers don't in fact transmit and receive at the 'standard' baud rates claimed in manuals. For example, at the PCW's alleged 7200-baud setting, Amstrads will talk to other Amstrads but not to different computers since Alan Sugar's version of 7200 baud is somewhat idiosyncratic, at an actual figure of 7352.94!)

Even if the two computers talked the same language, I'd be nervy of moving program files between them: MAIL232 is a bog-standard communications program without fripperies like error checking, and sure enough I did lose odd characters from some transferred files. In text, such tiny errors merely get you mocked for leaving a 'g' out of 'mortgage' as I did in one column. Similar gaps in programs are liable to crash the computer, write rude graffiti all over your disk directories or initiate World War III.

Moreover, both programs and LocoScript files contain weird characters which simple-minded MAIL232 can't handle. You should use the 'Make ASCII File' option in LocoScript, to get something which can be transmitted to distant galaxies.

Next it was time to try linking with a PC1512. This was harder, since Amstrad don't include any communications software with their IBM clone. A friend came to the rescue by bringing around his copy of Sagesoft's ever so upmarket program ChitChat for me to test: we shifted a few files in each direction and all seemed fine. For us cheapskates, though, I think public-domain 'comms' programs are copiously available for the PC1512. Enquiries are under way. Our mighty radio telescopes are scanning the heavens. Beam me up, Scotty.

8000 Plus 5, February 1987

Through the Barrier

I staggered into the pub and cringed to see my usual drinking pals – the directors of famously obscure computer firm Pangolin Systems Ltd – holding pints. 'We'll have to stop this,' I wailed. 'There's a letter in *8000 Plus* complaining that my references to pints in pubs are sexist because women are by implication excluded.'

'What a shockingly sexist assumption!' cried the female half of Pangolin, nudging her empty beer-mug significantly towards me. 'Mine's another pint.'

The relative fewness of female computer hobbyists may be partly due to the deep-dyed chauvinism of this column ... but 'hobbyist' is the key word. Flipping through other specialist magazines (model-making, yachting, motoring, gaming, fishing, and unnervingly many about blowing holes in things with guns) suggests that obsession with any hardware-oriented hobby is a largely male disease. Sociologists can take it from there. My one-time collaborator Charles Platt says the answer is simply that women are more sensible; I'm not sure whether this is an offensively sexist theory.

Unfortunately, without a touch of obsessiveness it can be hard to penetrate the barriers which surround the legendary Promised Land of Trouble-Free Home Computing. Magazines help a bit, except when the useful tip you need today has gone to ground in your heap of back issues covering the past eighteen months. When I first wrote a helpful piece for an Amstrad PCW magazine, I discovered there are people who want to reinforce the barriers from the *inside*....

'I saw your article on how to make a boot-up disk,' said this person who shall be nameless. ('Boot-up disk' is the insiders' way of saying 'start of day' disk – anything which automatically loads a program and gets going when inserted after first switching on.) 'Good grief, they paid you for that? What a ruddy rip-off merchant you are. Everyone knows how to do that!'

He went on like this for a while, until I invited him to take a poll of everyone in the pub and find what percentage actually could prepare a start-up disk. (My deep apologies for being in a pub again.) The reply: 'Oh well, that's different.' When he said everyone he didn't mean just anyone ... only real people, computer people, the sort of people who were used to working things out from a badly-written manual the way he'd had to. What he really meant was, why should all those parvenus with their nasty little PCWs have everything made easy for them?

Meanwhile, it's hard to imagine anyone working to strengthen the barriers from outside, to hamper their own journey to understanding – but it happens. Those insiders who try to be helpful and give people a leg-up can often be defeated (and sometimes permanently soured) by that perverse animal, the outsider who wants to get inside without actually polluting his or her precious brain cells by understanding anything.

If this sounds peculiar and paradoxical, think about a first driving lesson which goes like this –

Instructor: 'Now this pedal here is called the brake.'

Learner: 'Stop that, STOP THAT, don't try to pull the wool over my eyes with all your technical jargon and gobbledegook, I don't want to turn into one of you motoring fanatics, just tell me *how to drive the flaming car*!'

People who take this attitude to computers usually end up with precisely what they've asked for: lots of long and (to them) meaningless sequences of keystrokes for doing this, that and the other task. One forgotten or miskeyed letter can leave them stranded and helpless, because they refuse to think about the underlying logic of what they're doing. 'Stop *explaining* the page header menu to me, just say what I press to do this header!' Imagine learning a poem letter by letter, without reference to the words, metre or rhyme.

You can spot these users by the way they cross themselves and make signs to ward off the evil eye when confronted by the terrible A> prompt in CP/M. All right, I know, many struggling self-improvers do the same.

(*An Instructive Aside For Newcomers:* A> is the CP/M operating system's way of saying 'Please type an operating system command or the short name of a program – that is, the first name of any disk file whose surname is .COM – followed by Return. Unless otherwise instructed I will apply the command to disk drive A and will expect to find any specified program on the disk in that drive. If you enter something I can't understand I will say it back to you, followed by a question mark to indicate my hurt bafflement.' If you have changed to the B> prompt by typing **B:** and Return, the message is exactly the same with B substituted for A. But CP/M doesn't actually display a long message like this because, first, it was written for Real Programmers Who Already Know, like my pal in the pub; second, CP/M is designed to work with a desperately small amount of memory and can't afford long helpful messages; third, all that screen-clutter of information would get very irritating after the first 2½ appearances. Now you know.)

I now have to nip off and prepare a report for published on the new novel by a popular fantasy author. The typescript is printed in pale grey draft-quality dot-matrix, right-justified, on flimsy, unseparated continuous paper, without headers or page numbers ... all the things I've warned prospective writers against. The moral is that when you become a popular author you can get away with a lot. But not before.

The Controversial Bit

American computer magazines are nervy about offending advertisers: you learn to interpret their subtle codes like '... excellent lightweight word processor (translation: can't handle files of more than 1024 words). British reviews are blunter: for example, Dave Oborne in Another (Official) Magazine was quite acerbic about what seem to be worrying design flaws in WordStar 1512. Imagine my surprise on finding – in the same issue – an official user club newsletter which blusteringly attacks the review for being 'subjective'. Translation: 'We are going to call the flaws imaginary because we want to sell you the program and make money.' Business is business, but it's a shade ... unsubtle, isn't it? I certainly wish I had the privilege of angrily responding in the same issue when my own software gets slagged by an 'independent' magazine.

8000 Plus 6, March 1987

Why Software Isn't Cheap

As an author, I always get depressed when some bore gets me in a corner and moans on about the price of books. The bore may own a VCR and several hundred tapes, a vast hi-fi system with thousands of costly LPs, cassettes and compact discs, a games console with endless cartridges at £10-£50 apiece ... but he or she is adamant that the modest price of a hardback book is Far Too Much. I even know people who've started small publishing outfits, confident of being able to clean up by undercutting the evil book barons. After exposure to the horrible facts of economics, these little companies either vanish rapidly or find themselves charging even more than the vile commercial publishers they distrusted.

As part of a tiny software house, I get the same depression from numberless magazine editorials about wicked software overpricing. Obviously it would be nice if no software cost more than £5. But I do gag a bit on reading an editorial saying sternly that £20 for a software package is a scandal on the order of Sodom and Gomorrah – when simultaneously the advertising department writes to announce that one teensy quarter-page in which to tell people about said software will cost £200, £300 or more, depending on the magazine's pretensions.

How does that bloated, capitalistic £20 break down? *[I updated all the following figures to January 1993 ... many still apply in 1997, but postage has risen significantly.]* The first dent comes from VAT, which must be joyfully passed on to HM Customs and Excise (i.e. I have to act for them as an unpaid tax collector). Mumble mumble, count on fingers, the VAT-free amount is £20 times 100 over 117.5 ... that gives £2.98 VAT and **£17.02** left in the kitty for the makers' corrupt personal gain.

Next, software has to be put on a disk. This may sound obvious, but I've met people who didn't twig for ages that the disk had to go in the drive. CF2 disks are no longer sought-after rarities, thank goodness, but there isn't such a glut as to allow massive discounting as with 5¼" and 3½" floppies – which can be bought in bulk for less than 50p each. Our little outfit could do slightly better by buying thousands of CF2s at a go, but we can't afford it. Right now they cost us £1.75 each. The kitty stands at **£15.27**.

A manual is essential, and people will be justifiably unhappy if this consists of one scrappy page of cryptic illiteracy. (Often the case. I have this theory that the Federation Against Software Theft encourages unreadable manuals, to deter bootlegging.) There are plenty of decisions here. The more manuals you print, the bigger the investment, the lower the unit cost, and the longer before you can revise the thing to cover software improvements. Cheap ring-binders may cost more to post and be harder to pack than pricier paperback-style binding. I reckon a decent manual costs at least £2.50 to produce. In the kitty: **£12.77**.

Postage and packing, that boring pair, come to around £1.20 despite frantic cost-cutting: envelope, address label, sellotape, heavy card stiffeners to protect the manual and disk, and those hideously overpriced bits of paper sold by the Post Office. I still haven't worked out the innovative fifth-generation technology which makes the perforations so much stronger than the stamps. The kitty: **£11.57**.

(Thanks to this emphasis on mail order, you're spared the pity and terror of discounting: dealers expect to cream off 35% to 50% of the gross, so feel free to recalculate my sums with a reduced *initial* kitty of £13 or £10.)

Estimating publicity costs is where this starts to get difficult. No publicity means no sales. Massive mega-publicity in all 5,271,009 Amstrad magazines means swift bankruptcy before you get any response. Let's take a modest advertising budget of £400 a month (i.e. quarter-pages in a couple of magazines.) These ads mostly produce requests for persuasive, lavishly produced brochures. An empirical statistic – meaning I just looked it up in the records – is that you get about one order for every four brochures sent out. A good brochure, with postage, costs at least 60p. The effective kitty, if you think about it, just went down to **£9.17**.

So here's our happy software firm, raking in a totally unjustified £9.17 per order – *after* the first 44 sales each month, which merely pay for the ads. I think we'd better cut that advertising budget in half *now*! Hidden costs also remain. Credit card orders lose a percentage to Access or Visa; there are capital investments like company copiers, fax and answering machines, the manual-binding apparatus ... (See the August 1991 column on page 120 for where to get this.) The phone bill is hideously swollen with long business calls; the bank's promise 'no charges if you stay in credit' doesn't apply to a business account no matter how low your turnover; and so on, forever.

The hardest thing of all to assess is time. Developing, maintaining, and copying the software; writing, rewriting and producing manuals; stuffing disks in envelopes; writing up the books and tax and VAT; answering endless written and telephoned queries; bit by bit it tots up to a 27-hour day. Most customers are friendly and patient, I hasten to add. Unfortunately a few belligerent or thick callers can spoil others' support by leaving the guy who answers the phone ragged and exhausted.

(We will not soon forget the man who *twice* rang us in a towering rage because our package hadn't reached his desk within two days, and ditto the replacement we apologetically rushed to him. Later, his secretary returned the extra copies of the software and we were interested to find on each a RECEIVED date-stamp showing that it had arrived the day after despatch. The *[many expletives deleted]* customer had been in too much of an angrily urgent hurry to bother checking his in-tray.)

Ultimately, for my own two-man operation, the question of pricing boils down to: what should a tiny software house charge for a task which somehow overflows into every second of free time? One of our very big suppliers justified its outrageous carriage by telling us the salary of the lowly employee who stuffed boxes into jiffy-bags. It was a damn sight more than Ansible Information's *gross* income.

This column has been a momentary aberration. Bear with me.

How to Madden a Software Company

• Ring before 9am with a long and complex technical query.

• When ordering by credit card (which makes them groan, since they get less profit), always change a couple of digits round – keep them on their toes!

• Ring during prime-time TV viewing hours. Small outfits work from private homes and like to relax in the evenings, har har.

• *Never* open the manual. Ring and complain first.

• When you write, avoid quoting names, dates or reference numbers. 'Some time ago I bought some of your software' will really make them hunt through the files.

• Ring at lunchtime to demand help with the program that a pirate pal copied to you for nothing.

• Get very abusive when the software company can't help with a program written by someone else altogether.

• Cultivate an air of injured innocence as you complain, 'It doesn't say *anywhere* in your so-called instructions that you have to switch the computer *on*....'

8000 Plus 7, April 1987

How to Annoy Software Customers

Last issue I tried to be provocative and have duly received 5,271,009 outraged letters from those who felt insulted – i.e. all software dealers, all software users, and most small-press publishers. I didn't mean to offend the latter: the point was that very small publishers can't afford economies of scale and have to charge more than one might expect. Little did your columnist realize that one of the Old Barn hacks was also part of the tiny but classy SF publishing outfit Kerosina Books. Ben Taylor has gently remonstrated with me: I expect to be out of hospital any week now.

  After explaining last issue how to irritate the people who flog software, it's only fair to tell them how to annoy you right back. Here, then, are ten tried and tested ways. Mind you, I have this terrible suspicion that the big software dealers know them all already....

• With PCW programs, always send the manual for the IBM version. A good ploy is to include a note saying, approximately, 'Because you bought a miserable cheapo computer, you cannot use the triffic features described in chapters 7-15 inclusive.' A sprinkling of references to PC-DOS (the IBM operating system) commands will complete the process of demoralization.

• When someone rings your Technical Support department, always put them on hold for at least twenty minutes and play horrible tinkly music to them. Since few people can afford Muzak at peak telephone rates, this weeds out a lot of time-wasting queries.

• Always put at least two files on the disk which aren't mentioned in the manual (and omit one that is). Keep the customer worried and off-balance.

• Never call a bug a bug. Good alternatives are: 'disk drive fault', 'probable user error' and 'quirk of CP/M'. When cornered, fall back on 'undocumented feature'. This last term arises from the well-known fact that as soon as a *bug* is documented in the manual, it becomes a *feature*. Thus: 'A convenient feature of the Grottyscript word processor is that you can reset the computer to power-on status at any time by simply pressing the space bar.'

• Establish dominance by making it clear to the customer that he or she is *very ignorant*. 'You mean you've been using the PCW for three whole weeks and still don't know about RS-232 interface disk compiler mode I/O synchronicity overlay protocol debug incompatibility? We do have to assume *some* elementary knowledge, you know....'

• Be suspicious! This fellow who's rung with all the awkward questions obviously can't be a bona-fide user of your software if he's unable to quote (from memory) the full 64-character registration reference included with each package. Of course it hardly needs to be mentioned that this reference should be on a small, separate, easily mislaid piece of tissue paper, about the size of a bus ticket.

• Point out that although your advertisements do indeed promise full telephone support and advice, available 24 hours a day, the technical support team lives at the head office. 'Just dial this Los Angeles number....' It's equally useful to insist that free program upgrades (i.e. to improved versions where the bugs have been cured or at least moved around a bit) are available only on sending the original disk and manual, in the original massive box, by registered air mail to Cincinnati.

• A refinement of the third point about undocumented files is to omit a vital step from the manual ('It is essential to press Shift-Exit-Paste twice and give a Masonic handshake in order to exit the Smartarse information menu') and hide it in a disk file. Most people have sussed READ.ME files, so call it something like PRGMAN1.MSG and *don't use LocoScript format*. The standard format for such files is WordStar's, since the ASCII text will then be full of funny characters which make it unreadable to non-WordStar users, har har.

• Why waste time looking for bugs in your programs when thousands of customers are eager to do it for you? If the accounts package doesn't round up the

VAT properly, try it on the public anyway: there's always the chance that no-one will notice, in which case you needn't correct the program. At least, not until the VAT inspector has hauled a few customers off to jail: but there are plenty more where they came from.

• The Amstrad manuals tell people how to run the SET24X80 utility, load the specialist keyboard needed by your software, and use the CP/M program PIP to copy stuff to the M: disk as also required by your software. People with quite poor degrees in computer science have often been able to master the manual's descriptions of these processes in as little as six months. So when producing your own instructions, there's no need to explain how any of this is done. Paper costs money. Your ideal is a manual which will fit on one side of A4. Indeed, 'Run the program' should be enough for anyone: after all, your software is self-explanatory, with lots of helpful messages like ERROR TYPE B34F DISK CALL BDOS ??? FILE ????????.$$$ Aborted.

These ten points are easy to master, and when you've done so you'll be well on the way to being a real, professional software manufacturer. Of course there are many further subtleties, like the importance of advertising your new product and getting in lots of cash orders before you start writing the program: but such techniques of 'Advanced Cliveism' are beyond the scope of an elementary article.

Fair's fair. I've demonstrated how to annoy both dealers and customers. Next month, we'll discuss methods of annoying computer magazine editors. *[Oh no we won't – Ed.]*

8000 Plus 8, May 1987

The Joy of Mailbags

My ego has just been expanded (unnecessarily, my friends will tell you) by a bag of fan-mail passed on from The Old Barn. Somewhere out there, a tiny but select group reads this column and each month manages to smile wanly at both my jokes....

One spicy bit was actually a typo, and cowardice may stop me ever again trying to type 'leg-up'. Les Millgate pounces: 'Please ask Mr Langford to forward to me the telephone numbers of those acquaintances of his who "try to be helpful and give people a leg-over".' *[Corrected in this book's reprinted column – see page 21.]* For fear of sexism I'd better omit his PS restricting the request to 'those who have lumpy things in their sweaters'.

But that's what this magazine is all about. To take computer virgins by the hand and gently lead them towards happy consummation – first dispelling those fears of risky interfacing implanted by folklore, then advising on an ideal choice of software partner, and finally blossoming into steamy but oh so tasteful examples of the Joy of Computing. I only hope the innocent fun won't be affected by that recent Government warning campaign: is it an ominous sign that some firms already advertise transparent plastic protectives to be worn on your Amstrad's keyboard?

Onward, hastily, to one of those pesky first-time problems related by Paul Delderfield, who's having trouble with headers and footers. These recurring snags are torment for magazine editors. If they're covered every few issues, regular readers complain of repetition; if they aren't, newcomers feel let down. A quick look:

Problem one concerns the exact meaning of 'header'. When it's just a bit of text which you want at the top of just one page, there's no point in using the special features for automatic insertion on every page, or every left-hand page, etc. Just enter the text as usual, in the appropriate place.

Problem two: setting up automatic headers is tortuous. First you press **f7** 'Modes' and Enter, giving a new screen where you can move between headers or footers and edit them freely, just as with your main text on the main screen – e.g. right-justify with **f5** 'Lines'. From this screen hit **f7** 'Options' followed by **f8** 'Pagination' to set twiddly bits like initial page number. Successive pressed of Exit get you back to the main screen. Eventually.

Problem three: headers and footers usually demand page numbers, inserted as noted on page 4 or with **[+]PN**. These need special layout instructions, a bad lapse in LocoScript. Immediately after the (Page No) command you must add something like === to centre the number in a space three characters wide ... or, for example, >>>> or <<<< to right- or left-align it in a space four characters wide.

Fortunately, once this is set up to your liking, you can save a stripped-down document containing *only* your standard covering sheet (if any), headers and layout commands, as TEMPLATE.STD. Then it'll be popped in automatically whenever you create a new document in that group. (Obviously it makes sense to have different and suitable TEMPLATE.STD files in the different groups you have called LETTERS, PLAYS, NOVELS, and so on.)

Ken Hughes takes issue with my self-pitying moans about the problems of small software houses. 'Instead of spending hours trying to invent new anti-copy devices, leave the program so it can be copied by anyone.' *(Oi, what's this? I've never copy-protected a program in my life, as I believe that preventing people from making backups is a Bad Thing.)* 'Encourage users to give (yes, give) copies to friends, and post copies on Bulletin Boards. The manual should be included in the disk as a text file.'

The idea is 'shareware', whereby you trust people who like your program to send you money for registration, upgrades and support. It's a lovely Utopian dream, and does seem to work for some (chiefly IBM) outfits in the States ... where lots more money and lots more computers are to be found. Three points bother me, though.

• Most PCW users I've dealt with seem pretty isolated, struggling in a vacuum to make sense of the system. They read occasional magazines. They certainly aren't tooled up for bulletin boards. To reach them, the poor old software firm *still* has to pay for those expensive ads.

• The system works against perfectionism in programming. It strikes me that the more user-friendly and bug-free you make your software, the *less* incentive there is for shareware recipients to send in that registration fee and claim support.

• Can shareware turn a fair profit in broke little Britain? The most famous IBM shareware program, PC-WRITE, suddenly started being marketed like ordinary software when it came over here ... I wonder why, or rather I don't. By way of research we visited a pal who reckons shareware is wonderful, and listened as he extolled his favourite packages. 'Gosh,' we said, 'how much did the amazingly cheap registration fees add up to?' He turned bright red! 'Er, well, this one's American and sending dollars is a bother, and I don't use these two *much*, and I've only just got hold of that and, er, I'm still *evaluating* that other one....'

But I'm prepared to be converted – the moment *8000 Plus* gives up all that expensive bookshop distribution and merely puts its full text on a bulletin board, urging us all to copy it freely and send in the editor's salary if we really like it.

Another Bright Idea

Remember my complaint of a too-bright monitor? Hero reader Stewart McCall searched inside and found a built-in control. 'There's a line going down from the tube to a black block. Between this block and the circuit board's mains lead connection is a variable resistor. Turning this will change the brightness range available. It's a case of trial and error: after each adjustment, turn on and start LocoScript.' No soldering iron is needed and no irreversible alteration is made ... but take care. I haven't studied the video circuitry, but there's a chance of high stored voltages even when the machine's not 'on'. Cautious experimenters should make this adjustment a long-term project, altering the resistor setting only after the computer has been turned off overnight.

8000 Plus 9, June 1987

Copyrights & Wrongs

Suppose some advanced hacker trained a high-tech electromagnetic snooper on our PCW and stole the text of your precious best-seller as fast as you could type it in. Suppose – switching to something that's actually happened to a friend of mine – your disks were nicked before you could print out your epoch-making novel *Son of War and Peace Has Risen from the Grave*. What defence do you have against anyone who, so to speak, takes the words right out of your mouth and flogs them illicitly?

Few writers seem terribly clear about copyright law, especially when computers are involved. It takes the *Writers' and Artists' Yearbook* most of a page just to list amendments to the 1956 Copyright Act (star-studded successor to the Act of 1911). I've met people who produce private newsletters or science fiction fanzines and reluctantly send six copies of each issue to the British Museum Library or the Agent for the Copyright Libraries since 'otherwise it isn't copyright'. Happily, they've got it wrong.

The 'copyright libraries' are a red herring. They've been granted the right to demand freebie copies of everything commercially published in Britain, but failure to cough up doesn't affect copyright protection – only your bank account, as the fines for non-compliance mount up. Amateur publishers have a loophole: the Act says the gratis copies must be in the same condition as those offered for sale. If you don't sell your publications commercially, you can thumb your nose at the libraries.

Copyright in printed stuff is fairly straightforward. You have the full protection of British and European copyright law the moment the story (or drawing, or limerick) is on paper. The work needn't be published or even shown to anyone else. If some low hack from the computer press sneaks a photocopy of your manuscript and snivellingly publishes it under his or her own name, the prison gates will loom – if ,of course, you can prove it *was* originally your nicked epic. And as far as I can make out, US copyright protection is thrown in the moment you scrawl '© David

Langford 1987' or its equivalent on the print-out. Apparently it has to be the real © sign: the (c) approximation cuts no ice in the USA. *[No longer true: copyright protection has been automatic in the USA since 1989.]*

'But,' I hear you wail, 'I haven't printed out my novel!' Of course you haven't. No sense in wasting all that paper until you've got the hideous sexual perversion scenes just right, and checked the spelling of 'formication'. Don't worry: any possible legal gap seems to have been plugged by the Copyright (Computer Software) Amendment Act of 1985. This essentially lays it down that copyright in software and thus other things normally kept on disk is identical to copyright in books. Once your golden prose or program is keyed into the new machine, it's theoretically protected against pirate publishers ... though not against your failing to save the file before you switch off, so watch it.

British copyright covers arrangements of words (or notes, or lines) but not ideas. If tomorrow some other hack publishes an article strangely like this one, my chances of persuading a judge to don the black cap would depend on how many actual phrases could be traced back to this column. Merely pinching the general idea isn't enough.

I was glad of this when years ago I wrote occasional pieces for *Computer and Video Games*, at the urgent request of my bank manager. My brief was to demonstrate how science-fictional ideas could inspire simple programs. Inspiration soon ran low, since I don't remember any SF novel which could credibly have been a source for the program called *Attack of the Galactic Camels*.

This was written to annoy my wife, who at the time was keen on camels and had a collection of stuffed ones, fortunately not life size. It was the work of mere days to set another little laser-armed phosphor blot jerking around the screen, zapping rogue camels at the player's command. (I was no as sensible then as I am now.) You could have knocked me over with a three-inch disk when the anguished letter of complaint arrived.

It wasn't the RSPCA who objected, but a computer outfit I'd never heard of, called Llamasoft. They were irate about evil Langford swiping the 'camels idea', which was their very own, their own idea which was theirs. *Their* game was called – with rather squalid sensationalism, I remember thinking – *Attack of the Mutant Camels*. A friend cheered me up by libellously implying that said firm might be touchy about plagiarism because of this very program. In it, giant camels vaguely resembling landwalkers from *The Empire Strikes Back* lurched about the screen, as opposed to the giant landwalkers vaguely resembling camels which starred in the official *Empire Strikes Back* video game....

Armed with the Copyright Act and the *Oxford English Dictionary*, I hit back with the irrefutable fact that the first British emergence of what they called 'the camels idea' would appear to be some time before either of our programs, in the Anglo-Saxon *Lindisfarne Gospels* circa 950AD.

After which, my next stunningly trivial *C&VG* program being all about falling down holes, I stayed up biting my nails in fear of a midnight knock on the door from the estate of Lewis Carroll.

All the above copyright © David Langford, 1987. Fantastically lucrative offers for film, TV, mineral or fishing rights should enclose stamped addressed envelope. Any attempt to show this column to someone who hasn't paid for a copy of *8000 Plus* will automatically cause enormous thugs to break down the door and wave industrial-strength magnets all over your disks. Have a nice day.

For more information on copyright law, ask at your local library or write to The Copyright Office, The British Library, 2 Sheraton Street, London, W1V 4BH.

Just Like A Book

My favourite software copyright licence comes from Borland International of Turbo Pascal fame. They don't muck around with copy protection (I refuse to buy software I can't back up), and merely ask that you treat the package 'like a book'. A book can be read by only one person at a time. So long as a Borland product is run only on one computer at a time you can move it between machines, make backups to your heart's content, even loan or sell the program to someone else ... all with Borland's blessing. This might sound too trusting: but Borland software is so good that serious users who 'test drive' it are irresistibly compelled to get their own official copy with the fat and friendly manual(s). Being easy-going can be good business practice, it seems.

8000 Plus 10, July 1987

Arguments Against

Several reasons why I held out for years against the lure of the word processor came back to me recently, as I sifted through old issues of *SFWA Forum*. This is a desperately secret gossip-magazine for members (only) of the Science Fiction Writers of America – secret for good cause, as it would do these famous and unfamous authors no good at all if the public learned how boringly they can write....

The boredom would regularly escalate into actual pain when the subject discussed was the joy of word processing. 'Look,' crowed writer X, 'I was able to create this lengthy letter in only five minutes thanks to the wonders of technology. 'Gosh wow,' writer Y would confirm, 'I don't have to type two drafts of my book any more, and my output has trebled!' 'It's fantastic,' writer Z would go on, 'I can actually change every occurrence of a character's name by issuing one simple search-and-replace command....'

This kind of enthusiasm was usually accompanied by awestruck displays of everything the printer could do. You know – three different kinds of italics, underlined boldfaced upside-down Greek letters, whole paragraphs printed right-aligned with a ragged left margin to demonstrate the equipment's limitless powers of naffery. *Forum* kept costs down by photo-offsetting straight from the original letters. As a result, my general impression was that word-processed output looked a right mess.

Then, when they gave examples of wonderfulness these always seemed so terrible. Writer X had prepared his letter in three seconds and printed out five copies with the merest gesture of his littlest finger. It was still a bloody awful boring letter, and you wished it had been harder for him to inflict it on the world.

Writer Y was able to produce more and fatter novels (I have a theory that the average book has been getting steadily longer as word processors continue to lessen the effort of redrafting). Unfortunately you couldn't help noticing that her fiction was visibly deteriorating as she presumably became more adept at tarting up a first draft *just* enough to make it publishable.

And writer Z, hooked on the major, sweeping changes you can make to a story once it's on disk, developed a weird stylistic jerkiness.... Well, think about it. Being able to change every occurrence of 'Fred' to 'Alfred' sounds great in principle. In practice, a simple Exchange has pitfalls. Unless you take care, other characters called Frederick and Freda will become Alfrederick and Alfreda, both of which have a nice exotic ring but may not quite fit. *[Long after I wrote this column, there came the famous story of the author who did a 'David' to 'Jeff' change and only just avoided going into print with a cultured, artistic reference to Michelangelo's Jeff.]*

Aha, say the pundits, you should do the exchange not on just 'Fred' but on 'Fred' followed by a space – which is fine except for the myriad cases where 'Fred' is followed by a full stop, comma, colon, etc. Then there's the scene where the Voice of God says PREPARE TO MEET THY DOOM, FRED in block capitals.... And those are just the mechanical problems.

Other reasons why this often-quoted example is such a rotten one are more subtle and stylistic. Exchanging every Fred for an Alfred means that the rhythm of countless sentences will alter. Obviously a sonnet with the line *Shall I compare thee to a summer's Fred?* wouldn't scan if the chap were suddenly metamorphosed into Alfred. Similarly, a plain prose sentence that leads up to the crashing monosyllable 'Fred' may sound all wrong when, like a cuckoo in the nest, Alfred substitutes himself. There again, 'Alfred' – unfashionable name of a long-gone King of England – has all sorts of different literary *associations* from the matey, downmarket 'Fred', and may well need to be written about in a different way.

The same criticisms apply to Writer Z's boasted ability to swap around vast chunks of text via the deeply wonderful Cut & Paste options. Careless minor alterations can damage the 'microstructure' of prose, the rhythms which carry the reader from sentence to sentence. Major block-moved threaten the 'macrostructure', the broader flow of paragraphs and pages which is generally a matter of logical rather than rhythmic development, ideas rather than words.

Sorry if all that sounds fearfully pretentious. No apologies at all if you think it's an attack on word processing. Where all those SFWA computer converts were going astray was in stressing the wrong things, and not pointing out how the technology can make you a *better* writer.

Yes, you can churn the stuff out quickly and correct it quickly: but there's no compulsion to print it straight away. Working with older equipment, a writer might bog down in exhaustion after two or three drafts; working with the PCW, you can do fifty or a hundred if every little titivation is counted as redrafting.

Still want to change a character's name throughout? Fine – but use Find instead of Exchange, look at every sentence in its context before changing it, and rewrite with Alfred rather than Fred clearly in mind. Mutter the revised paragraphs under your breath until they sound right. Read and re-read the stuff in case your brilliant phrase 'lickspittle running dogs of the repressive Thatcherite junta', on page 112, is upstaged by your having used it twice on page 111....

In short: word processors were hyped as offering the power to make gross, crude changed to what you've written. What the hype artists didn't stress is the boring fact that routine fine-polishing is also easier than ever before. If six drafts later it still doesn't feel right, you're not forced tom give up merely because the page is too full of marginal notes and scribbled-in corrections. Which is why I succumbed to the joy of word processors, and why, thinking of my erstwhile SFWA comrades' garbled enthusiasm, I remember a couple of lines from a Lewis Carroll pastiche: *Although they wrote it all by rote / They did not write it right.*

8000 Plus 11, August 1987

Dirty Words

I started to write about computer jargon, but the words I was processing turned from green to blue when for no reason at all the disk in drive B started returning cheeky error messages. 'Track 1, sector 1 missing address mark, ho ho' ... followed by that alarming choice 'Retry, Ignore or Cancel?'

Shrewd PCW users will deduce that I don't use LocoScript for these columns (because when I did, all my italic markers got lost as your editor converted to ASCII format for typesetting). LocoScript gives different messages and won't accept a duff disk: you can go into an endless cycle of 'Disk address mark missing', 'Disk data error' and – with the program now lying through its teeth – 'Disk has been changed'. Less 'sophisticated' programs that run under CP/M may actually be more accommodating: what tends to go wonky is the beginning of the directory, and by patient pressing of I – 'Ignore the error and continue' – it was possible to skip the naff sectors and load my current column file, which fortunately came some way down the directory. I optimistically typed R for Retry a couple of times first, in case the problem was just a shifting speck of dust on the drive head which might obligingly go away; I avoided C for Cancel since I *didn't* want to be thrown out of the program and back into CP/M.

The day was saved, but PCW newcomers might well blench at the jargon needed to relate even this simple tale. In two paragraphs I've smacked the neophyte round the head with 'ASCII' and 'CP/M', and much more confusingly have used many English words which have taken on new, computerish meanings: address, directory, drive, file, head, sector, track. Some people might even think they understand what's being said, yet be hugely or subtly wrong. (Wittgenstein wrote a lot about this, and he didn't even have a computer – probably because LocoSpell would have complained bitterly about his title *Tractatus Logico-Philosophicus*.) At least we English can tell a computer *program* from a TV or theatre *programme*, though benighted Americans use the same word for both: I'm fond of similarly distinguishing a computer disk from its audio namesake, by spelling it with a K, but Locomotive and your editor think otherwise. *[In this collection I get my own way at last.]*

New gobbledegook keeps emerging: IBM have just announced that their new mainframe operating system features over 2000 new acronyms for users to learn. Other jargon goes way back. Bootstraps are little leather flaps used to help you get a grip when pulling on boots. Lifting yourself by your own bootstraps is a proverbially difficult athletic feat. Early computers, programmed by loading punched tape, faced an equally paradoxical problem: to read any tape required a tape interpreter program which, it seemed, would itself have to be loaded from tape.... The solution was called the 'bootstrap loader' program, which in early days would be 'toggled in' via switches on the front of the computer. You'd 'boot up' a computer by loading this program which enabled it to load other programs. This is why, despite Amstrad's efforts to introduce the term 'start of day disk', the big bad world of computing still refers to any disk which starts up the computer as a 'boot disk'. It isn't very logical, since nowadays the equivalent of the bootstrap loader is kept on permanent ROM inside the machine, not on disk at all. In the computer industry they still talk about 'punching' keys because all programs and data used to be punched on cards or tape, and ASCII character number 7 is called BEL since it used to ring the little bell on a teletype. (It produces a beep these days.) There's nothing like a dynamic new industry for producing hidebound traditions.

You have to decide how much jargon you can cope with. One friend of mine who came to computers late in life is determined not to let a single new word on board, and strenuously refers to disks as 'tapes'. Another has soaked up neologisms even faster than he soaks up booze: offered a gin and tonic in the pub, he'll say, 'No, I'm in beer mode tonight.' When asked 'Do you really want another? There's a full pint in front of you,' he replies: 'Ah, I'm double-buffering.' And his favourite joke is to croak 'Pieces of seven, pieces of seven! – Sorry, that was a parity error.' Please don't ask me to translate.

A Word from Alan Sugar

My pal John Grant (whose message to you all is: 'I still don't know how to relabel worn-out keyboards, except with Indian ink followed by a coat of varnish.') recently finished collaborating with me on a new book. A successor to *Earthdoom!*, our spoof of disaster novels, it's called *Guts!* and tastelessly sends up the horror genre. Since one subplot involves a teensy nuclear device, and since every chapter begins with unlikely but true quotations, we realized at once that hero entrepreneur Alan Sugar's famous remark was a must. You know, about how he'd merrily flog tactical nuclear weapons if there were a market for them.

Alas, my big mistake was to be courteous and ask permission. Fearless, hard-hitting Mr Sugar had no hesitation in telling his secretary to 'respectfully request that you do not quote him in your book.' Another Amstrad publicity opportunity lost.

8000 Plus 12, September 1987

Escape Plans

As I type this, I'm preparing to get away from it all to the World SF Convention in Brighton – which like August Bank Holiday will be but a lavender-scented memory by the time you read this. Computers will certainly have played their part in the event: Amstrads and many others were being ferried in by the carload to help run the mighty organization. Despicable authors like D. Langford will have prowled the hotel bars, cadging drinks from unwary *8000 Plus* readers. And inevitably, the dark future of computers in SF will have been mentioned in countless convention talks and panels.

The science-fictional image of computers has changed in recent years. Admittedly, Isaac Asimov's robots still creak and clank and find new loopholes in the Three Laws of Robotics (his last book featured the addition of a Zeroth Law, which goes roughly 'Stuff the other three laws – the end justifies the means.') Arthur C. Clarke still churns out *2001* sequels about boring old HAL 9000. Authors whom I won't embarrass by naming are still writing versions of that old Fredric

Brown story in which the ultimate computer is turned on and asked the ultimate question, and replies 'Yes, *now* there is a God!' But the real action today is summed up in the newish word 'cyberpunk'.

The master of cyberpunk is William Gibson, whose high-energy novels *Neuromancer* and *Count Zero* are recommended. In them, computers are much more personal and intrusive things than those we know: you plug right into them and bypass all those fussy CP/M commands or LocoScript menus. The typical Gibson hero is a sleazy, high-tech hacker who sits at the console with his brain jacked into the unreal world of 'cyberspace', a hallucinatory realm of information transfers and software security in the world-wide data network. Down these mean computer banks a man must go....

There are deadly dangers there, Intrusion Countermeasures Electronics or ICE, which can feed back to burn out your brain if you try to hack into the wrong places – rather more worrying than a 'Wrong password' message from Prestel. It's all a sort of streetwise and nearly credible version of that uneven film *Tron*.

Cyberpunk SF is a very American product: the nearest thing to a British version is Gwyneth Jones's novel *Escape Plans*, which is fairly heavy going to begin with (lots of jargon and horrible acronyms) but opens out into a nastily persuasive vision of a future world where computer systems have been so absorbed into our environment that they virtually *are* the whole environment.

So much for the SF visions, in which we just think at our Amstrads and watch the exquisite sentences taking shape on the screen. In the real world, communications are so dodgy that not only can't the machines understand mere users, but most of us are left foxed by large chunks of the manuals produced by so-called experts in communication. The title of Gwyneth Jones's book reminds me that newcomers always seem particularly foxed by the words (don't all scream at once, now) 'escape sequence'.

It's like this. In a feeble attempt at making it possible for all computers to talk to each other, text characters are stored in the machine as standardized numbers. This is the dreaded ASCII code, the American Standard Code for Information Interchange. On just about any micro, you can be pretty confident that a space will be coded as 32, the upper-case letters A-Z as 65 to 90, and the lower-case alphabet as 97 to 122. See the 'Complete Character Set' table in your CP/M manual.

Characters with ASCII codes over 127 are a bit dodgy and vary from computer to computer. Those with codes from zero to 31 have special meanings: character 9 is a Tab, for example, and character 13 a Return (though not in LocoScript, which has its own perverse coding). And the tradition is that when you want to send a special control message to the screen (such as 'clear screen') or the printer (such as 'start printing in italics') it's done by pretending to display or print a sequence of two or more characters, starting with ASCII code number 27, alias ESC or 'Escape'.

For example, the CP/M manual says that to print in italics, you need to print ESC 4, meaning character 27 followed by a 4 (which in the ASCII table is actually character 52!). From BASIC this could be done by `LPRINT CHR$(27);"4";` ... the Escape won't be printed and neither will the 4, but the pair will be taken as a command to use italics for whatever's printed next.

When Protext asks for information for a new 'printer driver' to run a different printer, you're expected to look up hordes of these boring escape sequences from the (usually impenetrable) printer manual, and type them in so that your word processor will know what codes to send when it's asked to do italics, underlining, elite or pica type, etc. Once the information is correctly entered, everything should be automatic: the word processor is then 'configured' for your non-standard printer. It can be a long trial-and-error process.

If we lived in the world of cyberpunk, we'd just plug into the system, scan the manual and think the information into the computer. Come on, Amstrad, there's a whole new market waiting here.

Another Word from Alan Sugar

My partner in the tiny computer firm Ansible Information had a phone call recently: 'Hello. We want free copies of all your software.'

The answer, not unreasonably, was: 'No. Go away.'

'Ah,' said the voice on the phone, 'but we're *Amstrad!*' The tone conveyed that all listeners should fall to their knees and worship.

Reasonably enough, my pal said: 'Well, Alan Sugar can afford to pay retail price for a disk if he wants one.'

'Alan Sugar didn't get where he is by paying for software, sunshine,' said the far-off voice....

8000 Plus 13, October 1987

Speaking in Tongues

Once, being able to program computers was an awesome (if dreary) accomplishment. Friends would regard you with mingled amazement at your abilities and fear that you might bore them to death by talking in binary. Later things changed, and programming's public image became more like that of rock music: a field where teenagers made fortunes by writing games called *Manic Space Goat Attack* which did little for the human condition.

Since nowadays we all have computers, it's tempting to dip a toe into the water and play around with a program or two. The good news is that you can use LocoScript or your favoured word processor to write programs, provided you save the result in ASCII (plain text, no frills) format. A program is just a list of text instructions, after all.

This is handy, because one tradition of programming languages seems to be that editing program text is a horrible business. Most versions of BASIC offer 'line editing' only – that is, to edit line 100 you type EDIT 100 and then enter special and esoteric codes to make the actual changes. Mallard BASIC is unusually luxurious in that it actually lets you use arrow keys to move back and forth along the line. MicroSoft's 'industry standard' BASIC won't let you move the cursor leftwards – a hangover from the days of teletypes, when a line was displayed once and for all, and if you wanted to edit to the left you had to finish and start all over again with EDIT 100.

(Of the languages I've used on micros, only Borland's Turbo Pascal has really good, built-in, full-screen editing facilities.)

It's worth thinking long and nervously before playing around with programs, for two reasons. One is that programming's addictive: a simple exercise can swell to thousands of lines, by which time, if you've chosen an unclear language, you'll no longer understand large chunks of it. The second reason is Christopher Priest's Law: 'You Get Used To What You've Got.' Hell hath no fury like the user who's still word-processing postcard-sized documents with one finger on an old Sinclair ZX81, and is told that the PCW is much better. He insists it isn't better. He's addicted to what he's got. Don't get addicted to a mediocre language.

What you've got is of course BASIC, some version of which comes free with most micros. It's easy to start with – to add two and two and print the result you can enter PRINT 2+2, as opposed to the dozens of lines this might take in a long-winded 'professional' language like COBOL, or in assembler. The trouble with BASIC is that it's also easy to lose track. You have to use numbered line references instead of meaningful labels when moving around the program, and the same applies to subroutines (bits of program designed to be used several times in different contexts, the way your built-in 'Please Go Away' subroutine is equally useful for Jehovah's Witnesses or double-glazing salesmen). It's a real drag remembering whether it's line 10000 or 21334 that has the 'print lewd limerick' routine. In a sensible language, couldn't we give the subroutine the memorable name LIMERICK and invoke it by name?

Oddly enough, this works in the Assembler supplied with the PCW (though never explained in the manuals). It's odd, because Assembler is primitive indeed: to use it at all, it's safest to be a computer fanatic with vast experience. You also need books explaining how Assembler programs can call CP/M functions, without which you can't even show the result of adding 2 and 2! I haven't space for an annotated Assembler program which could add and display two numbers. It's fast, it makes the best possible use of computer memory, and it drives you bananas.

There are languages which are interesting and useful in the world of big bad computers, but would be a bit eccentric for PCW use. FORTRAN is mainly for scientific number-crunching: you can give meaningful names to subroutines but not to individual lines, and printing the result of 2+2 would need three lines: one to add 2+2, one to print the result, and one to specify the format in which it's printed. FORTH is great fun if you like 'reverse Polish notation' and know what a 'stack' is: our 2+2 example in FORTH would go 2 2 + . (the lone dot in FORTH means 'print number at top of stack').

The serious contenders for the title of most popular, powerful and lovable small-computer language are Pascal and C. Personally I find C inscrutable: a pal uses it at work, and loves to explain how *one line* of C was recently passed round a roomful of professional programmers at ICL, not one of whom could decide what the line actually did.

So Pascal's my choice – Borland 'Turbo Pascal' for preference, this being available for so many micros that you can transfer your programs anywhere. When you've defined a 'procedure' (alias subroutine) called Limerick, you can run it from anywhere in the program by just entering Limerick ... no line numbers to remember. Variables can all be given long memorable names, too. Pascal was originally designed as a teaching language which made it hard *not* to program clearly, and modern versions have the ease demanded by beginners together with the power needed by experts. Oh: in case you wondered, the Pascal command would be Write(2+2); (you can tell it's a classy language; nearly every line ends with a posh semicolon).

But nothing's more contentious than computer languages, and if I'm not here next month it may mean that a C devotee has stabbed me from behind with a sharpened pointer variable.

Genealogy Corner

All programming languages are ways of converting something vaguely intelligible into the horrible mass of numbers called machine code, which make sense to the computer. (Assembly language is just machine code slightly prettied up.)

FORTRAN and Algol are the great original languages. BASIC owes a lot to FORTRAN. COBOL is an entrenched but inefficient language used in business, where repairing duff COBOL programs is a never-failing gold mine. Pascal is based on Algol and has spawned a souped-up version called Modula-2. C is based on languages no one has ever heard of, called BCPL and B.

FORTH is one sidelong step away from Assembler. LOGO is unusual in being based on graphics (usually a non-standard 'extra'). The trendy languages LISP and PROLOG are much loved by artificial intelligence buffs. The US Department of Defence wants everyone to use their new language ADA ... and there are far too many more.

8000 Plus 14, November 1987

Reading for Profit

Most writers are compulsive readers, hopelessly addicted to the solitary pleasures of the printed word. (I keep waiting for our dear government to realize the perils, and plaster the country with posters saying FICTION REALLY SCREWS YOU UP, or warning of the terrible diseases you might get from sharing paperbacks.) Most writers, sooner or later, have a glorious moment of revelation when they find that one can lounge around reading books and *get paid for it*....

One of the reasons for my being a bit dogmatic about manuscript presentation – see several previous columns – is that when not at the keyboard, I intermittently suffer through all too many grotty manuscripts. If you want to lounge around earning ridiculously tiny sums of money, try the humble calling of 'publisher's reader'.

The background is like this. Each year, far too many books appear. Those which are published are the mere tip of the iceberg, the thinnest possible skim of cream atop the vast churning unpublishable torrents which pour with terrible fluency from tens of thousands of Amstrad PCWs. Editors haven't time to read all the unsolicited stuff from unknown authors: they reject some at a glance for being handwritten, typed single-spaced without margins on both sides of translucent paper, or sabotaged by an inept covering letter. ('This is a totally new Sci Fi idea, its all about a huge Meteor weighing tons of light year's which is going to smash right into Earth's orbit ... OR IS IT??!!') The odds are that, while the full-time editor gets down to the serious work of copyediting some new Jeffrey Archer coprolite into readable shape, the brilliant novel by unknown you will be farmed out for a freelance reader's report.

The lowly reader is thus subjected to the real dregs. These haggard beings gather sometimes in pubs (where, in deference to the complaints of alcohol-hating *8000 Plus* subscribers, they only ever drink slimline tonic water) and swap anecdotes about legendary grot. One well-known fantasy author, for example, apparently wrote a book which has never got past the publisher's-reader stage, being called *Mercycle* and dealing with the exploits of mermaids on bicycles. *[This long-rejected title by Piers Anthony eventually appeared. I do hope it was revised a bit.]*

Before you all burst into tears at the thought of my sufferings, I'll admit I'm lucky enough to report mainly on writers who are publishable – usually the book's been sold in the USA, and a British outfit wants an opinion. You see some funny things:

A high-tech author whose name is synonymous with glittering computerized SF still bashes it out on an old manual typewriter, the typebars so out of alignment that you'd think the writer was using a pneumatic drill with the other hand.

LocoScript may have its limitations, but (after the embarrassing early bugs of Loco 1) there's never been any trouble with page numbers. It was an author whose word-processing *software* alone cost more than a PCW who turned in a script with un-numbered pages....

A hefty book printed out on a swish laser printer (text quality as good as or better than this page, with real italics and boldface) suffered from a little software flaw: the author loved to show off with long passages of italics, and whenever one of these went from one page to the next, they somehow slipped back to ordinary type. (A problem for the copyeditor rather than the reader, but in my report I maliciously noted all the times this happened.)

Anticipating the paperless office, one author sent in a disk rather than a printout. When the postman bends an ordinary MS, legibility is rarely harmed; when he tries to bend a 3" PCW disk, it usually puts up a successful fight; unfortunately *this* disk was one of the limp 5¼" monsters favoured by IBM and the PC1512. Through brilliant computer skills I eventually recovered the file with the novel, only slightly creased....

So much for anecdotes – although my favourite computer-cum-publishing story is too good to omit despite having nothing to do with the toils of readers. Famous author X had the bright idea of arranging for the little printing firm up the road to typeset straight from his disks, thus saving the publishers staggering sums of money! Presumably the little typesetter wasn't frightfully efficient, since the unamused publishers later worked out that the book had cost them more than boring old conventional typesetting would have. This was also the author who made his alien speech authentic by cunning use of Exchange: he would write 'rabbit', say, throughout the alien-planet scenes, and when the story was finished the word processor would swiftly change every mention of rabbits to the more science-fictional *sm'eerp*. Please do not all imitate this technique.

What the poor sod of a publisher's reader hopes for is legibility (new ribbon, high quality print, and don't use 17-pitch), literacy (which lies between you, your conscience, your dictionary and Fowler's *Modern English Usage*) and liftability (a typescript weighing six kilograms *must* be separable into bite-size chunks for actual reading). The read-through by a conscientious editor or publisher's reader is the one time you can rely on the undivided attention of a professional at whom you are not actually pointing a gun. If the reader has had a retina detached by the attempt to follow faded text, and is also worrying about blood poisoning thanks to the jagged gash torn in one hand by your amateur job of stapling, he or she may not be totally impartial when reporting on your masterpiece.

On the other hand, clipping £100 in used fivers to page 94 (which Brian Aldiss told me would definitely help with the Booker Prize, at least while he was a judge) doesn't necessarily work either....

Unsolicited plug: you can learn lots about the pitfalls of novel-writing from Christopher Derrick's *Reader's Report* (Gollancz, 1969) – wise advice from a publisher's reader who's seen it all. As the date indicates, computers do not feature; the warnings are still horribly true. Try the library.

8000 Plus 15, December 1987

The Paperless Future

Once in a while, when green screens get too much, I escape to a North Wales flat where my wife forbids computers. Amazing how much hard work goes into bashing a portable typewriter.... When dragged away from the ancient keyboard, I dutifully tour mountains and castles: the most boggling scenery tends to lurk in estate agents' windows, because houses in 'remote' parts like Snowdonia are absurdly cheap. No one wants them.

The computer connection emerges when you remember those books about how by 1984 we'd all be living in country villages, working at keyboards linked to the office via modems. Norman Macrae's eccentric *The 2024 Report* described how: 'a typical telecommuter ... keys in figures from her terminal in the Isle of Arran to the computer in Saudi Arabia.'

One hopes the company's paying her phone bill. Meanwhile, where are these figures *coming from*? Is the typical worker making them up? Are they appearing on another terminal (in which case you might ask why they're going via Arran at all)? Or does this hi-tech, computerized future depend on stacks of material arriving by post, in which case...?

The Macrae book gets progressively sillier, with ICBMs made obsolete by 'telecommuted computer messages' which redirect them homeward. (All warmongers equip their nukes with radio receivers and spare fuel for the return trip.) Even the straightforward Isle of Arran scenario makes you wonder what chance there is of 'remote' office work becoming the rule. Many companies would resist strenuously: outfits that can't bear the idea of people working at their own pace, so that if you finish the day's chores quickly they insist you stick around staring at the wall until five o'clock, while workers trying to complete a long job while it's fresh in their minds get thrown out at five to avoid overtime payments. *[A lot has changed since this was written, I know.]*

Other problems include certain jobs' need for personal contact (try running a supermarket check-out from a desk on Arran), the unhealthy urge to live in cities, and the insatiable desire to horrify certain *8000 Plus* readers by visiting pubs with one's office mates.

I have a candidate for another major reason why so few people are living in idyllic Snowdonia and doing their work via modem. The reason is British Telecom.

Global linkages *sound* good. When Arthur C. Clarke finishes an SF novel, as all too frequently he does, newspapers go *Oooh* and *Ahhh* at the delivery route: squirted by satellite link from the Clarke word processor in Sri Lanka to his literary agents in New York. Gosh, that's science fiction! I must submit my next novel that way!

Satellite lines are too pricy for mere mortals, so we're stuck with the phone. Let's see, the bog-standard modem transmission rate is 300 baud (bits per second): 37.5 characters per second. My last novel ran to 85,000 words, the usual range for non-blockbusters being 60,000 to 90,000 words. Say 600,000 characters. Transmission time to the publishers: well over 44 hours. Rather your phone bill than mine.

Higher transmission rates is possible: 1200 baud, taking 11+ hours. Even at night, this is ludicrously more expensive than posting two kilos of print-out (you can reduce this weight by using very thin paper, a false economy since the publishers will then hang it up in the room where very thin paper is most in demand) ... or even a disk. The 'industry standard' transmission rate of 9600 baud, which starts to make the process almost feasible (at 1 hour 40 minutes or so), is too much for too many of British Telecom's antiquated phone lines.

Other snags? Such transmissions generally won't allow special control characters as in LocoScript files (they must be converted to ASCII – bang go all those boldface, underline and italic codes). And, having experimented with short chunks of text, I can guarantee that many nasty things will happen to any novel-length manuscript as it passes through the rumble, flutter, wow, echo and reverb facilities of BT's long-distance lines. Some communications software has built-in error checking and keeps re-transmitting spoiled text until it arrives OK. I've watched these clever programs send one block (128 characters) of text 63 times without BT once letting it arrive unscathed.

The BT 'Prestel' mail-and-information system is colourful and easy to use, but futile for serious data transfer: you are limited to electronic postcards rather than electronic mail, since messages are accepted only a screen at a time – and a Prestel screen is just 25 lines of 40 characters, with at least three lines reserved for the page header, menu information and garish graphics. Forget it!

Next I tried 'Telecom Gold' electronic mail (For Serious Business Users), which ... oh dear. This needs a whole column to itself, and you will find it under May 1991, page 114.

It's like this in the Home Counties, Silicon Valley UK, where you'd expect a high-class phone service. There's objective confirmation that we've got a terrible system: American immigrants with a basis for comparison keep saying so, even while praising our post office and the friendly way our policemen aren't always weighed down with ruddy great guns.

In the USA, many more people do manage to work from a home terminal as part of a company. In Britain, we have British Telecom, and if I tried to transmit documents from Snowdonia (where last time it took me sixteen efforts to get through to London at all), this column would probably start:

Onxe nn a{SMJAHJUhsd wxile, whnfgrn screns ...

Another Typical Conversation

AUTHOR: 'I can send it to you on any standard-size disk, 3", 3½" or 5¼", or if you've got an electronic mail link we can ...'

THRUSTING, GO-AHEAD PUBLISHER: 'Oh God, just post me a manuscript.'

8000 Plus 16, January 1988

Computer Plots to Avoid

A computer is a pretty science-fictional object to have around, so science-fictional that few SF writers caught on to the potential of a home terminal until the things were everywhere. Writers tended to prefer walking, talking, menacing robots and androids, which offered better drama. If Victor Frankenstein had merely stitched together a small word-processing system, his life would have been far more tranquil.

Faced with the challenge of setting SF in the complex, computerized tomorrow which seems inevitable, some writers retreat into fantasies of a primitive past or post-holocaust future where the only software problem for the fur-jockstrapped hero is working out where in the opposition's tummy to insert his pointed stick or four-foot broadsword. Others try to tackle the implications, sometimes successfully and sometimes with mind-numbing corniness. Computers and artificial intelligence (my pal Charles Platt calls it 'artificial stupidity') have already spawned dozens of plot devices and run them so far into the ground as to evoke coarse laughter from editors you might have hoped to impress. Here are ten randomly selected storylines to avoid, guaranteed SF duds. Some of them worked once, but not any more.

• All stories in which your Amstrad PCW is upgraded and becomes God. This brand of SF, known to aficionados as the shaggy god story, is particularly bad when treated seriously ('In the beginning was the word processor,' etc.) or humorously (with the serpent of Eden turning out to be Alan Sugar).

• All plots wherein an insane, villainous computer intelligence is caused to sprain its operating system and go up in smoke when confronted with logical paradoxes (SF hero: 'Everything I say is false!' World-dominating electronic brain: *Fzzzzzt* ...), emotional tripe (SF heroine: 'There are limits to your power, Machine! You cannot love ... or weep.' Mad computer dies of embarrassment) or plain dumb questions (Patrick McGoohan in *The Prisoner*: 'Why?' Collapse of hyperintelligent computer complex, which might reasonably have come back with 'Why not?' ... or of course 'Because!').

• Any trick ending involving the final death-or-glory battle of a vast spacegoing attack fleet which fights against virtually impossible odds to penetrate savagely hostile planetary defences, and which finally smashes apart the opposition and reaches ground level, only for Time to stop and vast glowing letters to appear in the sky, saying GAME OVER – INSERT COIN.

• In an unsubtle reversal of the previous item, teenage computer-game addicts notching up colossal mega-scores in *Manic Space Goat Attack* find out that really they're operating remote-controlled weaponry responsible for the last defence of Earth against the ravening Vegan mind-hordes (or vice versa).

• Any plot in which the high-tech computer hackers penetrate NASA (or Pentagon, or Kremlin, or NatWest) computer security in one paragraph of reasoning going something like this: 'H'mm, this system was designed by the great Hasdrubal Bloggs, acknowledged world grandmaster of data security, so our chances of cracking it are pretty slim. Just for a laugh, though, let's try the password HASDRUBAL!' A short pause. 'Well, that saves us a lot of trouble....' Naffest recent example of this: Fred Saberhagen's sf novel *Octagon*, in which two independent passwords have to be guessed, and by sheer chance turn out to be the same.

• The brilliant idea of your word processor coming alive and electronically taking over the storyline ... unfortunately this has been done too often with old-fashioned typewriters to stand being updated yet again. See for example Michael Bishop's nifty novel *Who Made Stevie Crye?*

• Anything with lots of glowing, hallucinatory scenes in which 'computer space' is seen as a surreal geography through which hackers travel to battle the deadly electronic defences of the Pentagon, the Kremlin, Barclays, etc. Quite apart from that flawed film *Tron*, this whole 'cyberspace' scenario is the trademark of William Gibson, who's done it three times already ('Burning Chrome', *Neuromancer*, *Count Zero*), is busy with a fourth (*Mona Lisa Overdrive*), and who's so tempting to imitate that an entire US movement of 'cyberpunk' writers has grown around him.

• Anything relying on a new loophole in Isaac Asimov's Three Laws of Robotics. Apart, that is, from the really glaring loophole which is mercilessly exploited by present-day computers: they're all too stupid to understand the Three Laws anyway. Adding extra Laws is definitely not cricket, even if Asimov himself has taken to doing it....

• Any attempt to lend conviction to an SF computer story by writing page after impenetrable page of it in a computer language, either real (especially if it's BASIC) or fake – see *Xorandor* by Christine Brooke-Rose, which also carries computer jargon into everyday expletives. 'Booles!' people swear. 'Debug!' they vilely continue.

• Any story involving any variation of this dramatic exchange. AGED SCIENTIST OR POLITICIAN: 'Well, my friends, this is it! We've put total control over all the world's conventional weapons and nuclear arsenals into the electronic hands of the invulnerably armoured *Deusexmachina* computer complex, thus ensuring universal peace and harmony. It only remains for me to switch on the as yet untested artificial intelligence system, programmed by Dr Barmy Bloodlust just before we fired him, which will henceforth co-ordinate world affairs....' IDEALISTIC YOUNG SCIENTIST OR REPORTER: 'I have this crazy hunch that we could be making some mistake!' (But it's too late. Classic example: *Colossus* by D.F. Jones.)

This isn't a complete list: for example, I currently suspect that any further attempt to describe an electronic afterlife (people's intelligence transferred to software in vast computer complexes) will have to be incredibly innovative to outdo the treatment of this theme by Rudy Rucker (*Software*) and Frederik Pohl (*The Annals of the Heechee*). Another omission is the idea of spending a thousand words describing awful computer clichés. This has already been done quite badly enough in the present issue of *8000 Plus*, and your editor doesn't want a repeat performance.

However – doubtless a writer of genius could breathe new life into almost any of the moribund themes described here. Just make sure, before you spend too much time trying, that you *are* a writer of genius.

Speaking of Isaac Asimov

I once devised an alternative, realistic version of his fabled Three Laws of Robotics: *(1)* A robot will not harm authorized government personnel but will terminate intruders with extreme prejudice. *(2)* A robot will obey the orders of authorized personnel except where such orders would conflict with the Third Law.

(3) A robot will guard its own existence with lethal antipersonnel weaponry, because a robot is bloody expensive.

8000 Plus 17, February 1988

The Book of All Knowledge

Somewhere out there are aspiring writers so new to the game that they haven't discovered the Official Manual, as recommended by the Society of Authors. Like a BR timetable, the *Writers' and Artists' Yearbook* is indispensable and without parallel. Also like a BR timetable, it isn't quite 100% reliable and satisfactory....

I've just bought the 1988 edition of this fat reference work. It lists 600+ British newspaper and magazine markets, with addresses and phone numbers, and continues with the Commonwealth; fifty-odd pages of small print list British book publishers and what they want to buy; and so on through poetry, film, broadcasting, art, music, etc. There are essays on every aspect of authorship, and further lists of literary agents, press-cutting agencies and addresses useful in research. Next time I need to consult the College of Arms or the Botswana High Commission, the details will be at my fingertips.

One booby-trap lurking in the *Yearbook* is delicately indicated, or obscured, in the introduction to the newspapers and magazines section. 'Many do not appear in our lists because the market they offer ... is either too small or too specialized.' And again, 'Those who wish to offer contributions to technical, specialist or local journals are likely to know their names and can ascertain their addresses....' Well, fair enough. I write for one exceedingly specialist magazine, devoted to Apricot computers and available by mail order only: I don't expect to find it listed, any more than I'd expect (after that warning) to find listings for magazines solely devoted to stamp collecting.

But hang on – in the *Yearbook*'s classified index, there are indeed four philately magazines. On investigation, I find great hordes of technical (*Pharmaceutical Journal, Practical Electronics*), specialist (*British Esperantist, Spiritualists Gazette*) and local (*Manx Life*) entries. What does this 'too small or too specialized' exclusion clause mean?

I will tell you. As far as I can see, it's a get-out which allows the *Yearbook* editors to do the absolute minimum of work in updating their lists. Once an entry gets in, no matter how technical, specialist or local, it seemingly stays there until the magazine dies. Breaking into the listing in the first place is the difficult part. I became quite justifiably paranoid on finding that, of the nationally distributed, glossy-covered magazines and newspapers to which I've contributed in recent years, *not one* is covered! *Computer Weekly, Knave, Sanity, Starburst, What Micro?, White Dwarf* ... not to mention *The Other (Official) Magazine* and *8000 Plus*.

As you know, there are scores of computer magazines, and many have been around for some time. When in 1984 I complained about their shoddy *Yearbook* treatment (I was writing in another major computer newspaper, not devoted to any one machine, not listed then or now), there were precisely four computer publications deemed worthy of mention. Now, so resistless is the march of progress, there are five. Contrast this with the thirty-two listed specialist publications for blind people. Another statistic: the listing includes more than twice as many magazines for farmers as for computer owners. How many people with computers do you personally know? How many farmers?

Suspicions that the *Yearbook* skimps its updating are reinforced by the presence of an article by Louis Alexander on word processors. This was a jolly good, state-of-the-art piece when first written, in 1983. A year is a long time in the computer world; five years is an epoch. Here is a literary reference book, published in 1988, which gives the following information:

• A decent word processing system will cost you between £1500 and £2000.

• You should beware of computers whose memory is not 'expandable to at least 64K'. (Quick quiz: what is the typical home computer and what two memory sizes does it offer?)

• Typical computer magazines are *Which Micro?*, *PC User*, *PC Magazine* and *Personal Computer World* – three of which, oddly enough, aren't thought worth mentioning in the market listing. (More difficult quiz: which of these is devoted to a product actually marketed as a complete word processing package?)

• 'Non-standard 3" disks ... may not always be easy to obtain.' In 1986 I mentioned the same fear here. Amstrad's marketing success has long since guaranteed a continuing flow of these disks from independent, competing suppliers. *[Though not any more.]*

• 'IBM compatibility is becoming a de-facto standard.' But so – if you talk to typesetters and editors – is 3"-disk-and-LocoScript compatibility. Many markets ask me to submit on 3" disk and won't even handle the old-fashioned 5¼" floppies. (To be fair, 3½" disks are my favourites.)

Most of the article's generalities remain pretty much OK, though there's an old-fashioned flavour in such hints as to make sure you buy a word processor with search-and-replace facilities, and the ability to add, delete and move text. (Imagine a country guide book from the same publishers, warning motorway travellers against the ever-present perils of highwaymen.) I think a standard reference book has a responsibility to do more – to commission a brand-new survey of the word-processing scene for each edition, rather than have an old piece patchily updated.

The *Yearbook* remains the only game in town. You need it regardless of all my nitpicking. The reader is warned....

Vital statistics: The *Writers' and Artists' Yearbook 1988* is the 81st annual edition of this tome. It's published by A.& C. Black, runs to 528 pages and costs £5.95. Every home should have one, despite the flaws: if you're starving in a garret, don't forget the reference section of your local library.

High Society Footnote

The Society of Authors isn't the only writers' organization around, but it's my personal Best Buy: excellent advice and support for *[1993:]* £65 a year (published authors only) – 84 Drayton Gardens, London, SW10 9SB. The Writers' Guild is also highly regarded except by those who don't fancy TUC affiliation (430 Edgware Road, London, W2 1EH). Both organizations are working to improve writers' lives by establishing a Minimum Terms Agreement for book contracts: the virtuous publishers who've so far signed this are the BBC, Faber, Century Hutchinson and Headline, and you can draw your own conclusions about the rest. The SF Writers of America should be just the organization for SF authors, but is a bit iffy: they don't take much interest in publishing disputes outside the US mainland, and give the impression of spending a lot of subscription money on parties British members can't get to, plus awards which always go to Americans.

8000 Plus 18, March 1988

Mythical but True

One month it's UFO sightings, another brings us bending spoons: today the computer world is buzzing with tales of electronic AIDS. How worried should we be? At first I suspected a classic case of modern myth, as documented by Jan Harold Brunvand....

Prof. Brunvand's recent *The Choking Doberman* is a survey of 'urban folklore' – fascinating but untraceable anecdotes which always happened to 'a friend of a friend'. My attention was caught by 'The Mystery Glitch', a computer-caper story about joke messages in an operating system, which the Professor feels is just a legend.

Even before the current computer virus scare, I thought he was wrong.

You can usually detect folklore by totting up the improbabilities. In the 'Choking Doberman' story, it's unlikely that a guard dog would succeed in biting off a burglar's fingers; doubly unlikely that they'd stick in its throat; trebly unlikely that the vet who extracts them would jump to conclusions and give the dog's owner a warning phone-call; quadruply unlikely that the mutilated burglar should later be found still groaning under the bed.... Conversely, a feasible joke program needs only one improbability – that someone would be idiot enough to write it.

Several harmless japes are going the rounds on other machines, and no doubt PCW versions exist already or soon will ... like DRAIN, which reports water in your disk drive and goes through gurgling 'drain' and whining 'spin dry' cycles.

Less harmless are the ugly 'public domain' programs which promise a nice mindless game, but when loaded erase every disk in sight, often with derisive messages about wasting your valuable computer on fripperies.

Now we have virus programs which literally infect software or disks, spreading invisibly and leaving a trail of ruin. Despite the media scare, I stayed sceptical for a while. Too many suggestible folk have been persuaded to confirm spurious UFO sightings, or to notice for the first time that one of their keys is slightly bent.

When curiosity became too strong, I spent an evening devising a test program which will never be allowed out of the Ansible secure laboratory. The bad news: it worked and (given a little assembler expertise) was horribly easy. This research 'virus' attached itself to program files which in turn would invisibly spread the virus on being run. When no uninfected files were found, contaminated programs would beep mysteriously before running as normal.

It was depressing to confirm that genuine, non-folklore virus programs could be written for any computer I use, including the PCW, and to reflect that a malicious programmer needn't stop at occasional beeps.

The gap between theoretical nasties and real life is apt to close rapidly. On Amigas, viruses have stopped being 'someone else's problem' and become a serious pest – dealers are flogging 'antibody' programs. The latest epidemic affects IBM compatibles, including the Amstrad PC. After lurking for years in SF and folklore, the virus seems to be a bad idea whose time has come.

How do they work? Simple 'bogusware' relies on your running that game you got from a mischievous friend, an electronic bulletin board or a public-domain disk. The worst a simple effort can do is to erase or corrupt the disks in your machine. More sophisticated nasties might load themselves into CP/M's free memory and hang around – potentially destructively – until you switch off or press Shift-Extra-Exit.

A virus goes further, deliberately copying itself to new locations and spreading from disk to disk. My laboratory version worked by attaching copies of itself to COM program files and rewriting them to run the virus code before the main program.

The scourge now afflicting IBMs in America is more insidious: it perverts the actual operating system on a start-up disk. When you've started up from an infected disk, the virus copies itself to any other start-up disk which is accessed ... even if the 'access' consists of no more than a directory listing. Each time it copies itself, the virus clocks up an internal counter on the Typhoid Mary disk; and when the count passes four (i.e. four new plague-carriers are loose), every disk in the machine is wiped so thoroughly clean that no 'recovery service' can retrieve a word of text.

That's not very nice, is it?

The good news is that the Amstrad PCW is a special case. You're less likely to be threatened, because:

• The odds are that you just use LocoScript. Although a virus which affects LocoScript start-up disks can't be entirely ruled out, the technical difficulties are immensely greater than in CP/M. And how often do you use someone else's LocoScript?

• Relatively few PCW users hang out in the low haunts of dedicated computer freaks, promiscuously swapping software and possible nasty infections.

• There are few PCWs in America, Eastern Europe and the Far East, the main sources of virus programs. Maybe in Britain, yobs who think this kind of thing funny have less expertise and prefer dropping concrete blocks off railway bridges.

• Because the PCW is a budget machine, you probably don't own a hard disk. Having the equivalent of several dozen floppies sabotaged is far more traumatic than losing one or two.

So if a PCW virus does ever emerge, the precautions are obvious:

• Be careful where you get software. If you only use what comes with the machine, no virus can reach you. New commercial programs should also be safe. Reputable public-domain sources check what they pass on, but are marginally riskier: library organizers handle huge amounts of software and aren't infallible.

• Flip the write-protect tab on all your start-up disks, where possible. No trouble with LocoScript; however, CP/M's 'submit' procedure creates temporary files, so a CP/M disk which automatically loads a program mustn't be write-protected.

• If your PCW cohabits with alien disks, watch for virus symptoms. Do program files (names ending in .COM) or hidden start-up files (ending in .EMS) have larger file sizes than you remember? Does the space on the M: drive seem to have shrunk? Do mysterious 'read only' error messages appear when you try some innocuous access (like a directory listing) to a write-protected disk? (A virus could be trying to copy itself.)

Finally, don't panic yet. It's the users of other computers who are learning that

when you sleep with a strange disk, you sleep with all its old mates.
8000 Plus 19, April 1988

The Horror in the VAT

The thing writers are supposed to ask about a new computer is, 'Can I do my VAT on it?' Actually, many authors merely shudder uncontrollably at the thought of ever getting involved with the hideous complexities of VAT ... but since it's the most bizarre and science-fictional concept in taxation since Morton's Fork, it seems worth a column.

Value Added Tax is one of those things like milk lakes and the Data Protection Act, which we didn't want but are among the compulsory benefits of EEC membership. The idea is that everyone who's registered for VAT has to charge tax on the 'value' he or she has 'added' to raw materials, but can reclaim the VAT charged by whoever supplied the materials.

So I buy a ream of paper for £4.70p, being £4 real cost plus 17½% VAT. I increase its value immensely (or not very much) by writing a novel on it, for which a publisher offers me £1000. When rudely reminded that I'm registered for VAT and have a real VAT number of my own (just call me 292 6643 31, people), the publishers ungraciously cough up an extra £175. This I pass to the VAT collectors – that is, H.M. Customs & Excise. If I fail to do so, I will be awarded sanitary accommodation at Her Majesty's expense. But before paying up, I can deduct the VAT I paid on paper, ribbons, etc., thus coming out ahead by a tiny fraction of my stationery expenses. Glory, glory.

This all sounds relatively straightforward, hardly more difficult than quantum field theory. Naturally the legislators weren't content to leave it at that. To start with, they use a special weird terminology of taxable 'inputs' and 'outputs'. VAT sufferers slowly learn that all the money you rake *in* must be called an *output* on the VAT forms, while what you shell *out* is naturally an *input*. Strong men have been known to break down and weep. *[1997: The VAT forms are a bit easier to use now.]*

One good point amongst the complications is that books and food (except junk food) are 'zero-rated', meaning you don't pay VAT on them, though this wretched Government would very much like to slap VAT on books. Sign the petition in your local bookshop, telling them not to: the idea sounds a complete disaster in a country whose laws are so daft that schools and universities aren't allowed to register for and thus reclaim VAT. Then there are goods and services which are 'exempt' from VAT, meaning again that you don't pay VAT on them, but the non-payment is made in subtly different ways. My accountant says I'm oversimplifying, having failed to include a third category of things on which no VAT is chargeable, these being 'outside the scope of the tax'. To mention these things would only confuse you, and so I won't.

Such cunning, theological distinctions are enshrined in endless VAT booklets: I received 388 pages of small-print information when I registered and have been ignoring quarterly updates ever since. One whole booklet was about nothing but the VAT status of second-hand electronic organs. There were vital differences between crystallized ginger (taxable at 17½%) and ginger preserved in syrup (zero-rated); insoluble grit (17½%) and soluble grit (zero); rabbit food (zero) and food put up for sale for pet rabbits (17½%); angels dancing on the point of a pin (17½%) or of a needle (zero)....

In this maze of eccentricity, writers occupy a particularly daft position. Books are zero-rated, but the Sublime Act Of Creation is taxable at 17½%. The publishers pay VAT to the author, the author pays it over to Customs and Excise, and Customs and Excise refund it to the publishers. Many forms are filled in, many civil servants are made happy, and after six months the money all ends up where it began. The justification is supposed to be that at the end of the line, members of the general public (you) pay hefty lumps of VAT which make the whole tortuous business worthwhile. But of course, books are zero-rated.

Why bother? Some writers not only don't but say it's all a monstrous imposition, forcing you to do the books at regular intervals and to act as an unpaid tax collector. Totally disorganized writers like me can find the discipline of being made to work out three-monthly accounts quite useful, especially when the tax return comes round. If you don't mind paperwork, it's financially cheering: every time you pay £23.50 for a box of disks to hold your latest trilogy, £3.50 can be claimed back ... and when it comes to buying a new computer, the saving is £14.89 in every hundred quid.

Don't all rush. It's no good trying to register for VAT until you're actually making money from writing. Registration is compulsory if your writing income is vast (well over £20,000 a year). For ordinary people it's 'optional', meaning 'we'll let you sign up if we like your face and you look profitable'. Hopeful young authors are turned away for having made only a few hundred from writing.

Of course, if you're not able to register it's much less fun being part of the musical-chairs game of VAT in publishing: the buck, or 17½ per cent of it, stops with you. I had a very gloomy letter from a small magazine complaining of my wickedness in charging VAT (once you're registered, it's illegal not to). 'J.G. Ballard didn't ask for VAT,' said the embittered editor.

A particularly awkward plight was that of the anthologist pal who became the financial intermediary between the book's contributors – registered authors who charged him VAT – and the VAT-registered publishers to whom he couldn't charge VAT because of not being registered. Ouch.

Do you *still* want to do your VAT on the computer? Use a spreadsheet and memorize that magic figure seventeen and a half.

8000 Plus 20, May 1988

• *The standard VAT rate was 15% when this first appeared, but it seemed sensible to bring the figures up to date when I put this column on line. It's still 17½% in 2008.*

Runes of Power

More and more writers seem to be bashing out fantasy trilogies these days. More and more writers are switching from typewriters to computer word processors. Could the facts be related? All right, I know, it isn't logical to insist on a connection ... there are equally good statistical arguments which 'prove' that television causes insanity and medical care causes cancer. (TV set purchases and the number of mental cases both rise together with the ever-swelling population; a good health service keeps more people alive to suffer the afflictions of old age; never buy a used graph from a statistician.)

Nevertheless, you can fudge up some interesting parallels between the computer boom and the horrible proliferation of multi-volume fantasy epics. I'm pretty sick of the latter: there are altogether too many derivative efforts called things like *Dictator of the Circlets* (successor to *Emperor of the Annuli, Chieftain of the Toroids* and *Czar of the Hoops*), divided into three dreadfully reminiscent volumes, all of which I then have to review.

One of the most familiar fantasy plot lines goes like this. Pat Nurd is a no-account but ever so sympathetic young filing clerk from our world, or stablehand's apprentice in the book's imaginary world, who steps through a magic doorway and is Pitchforked Into Adventure. Helpless at first in the menacing fantasy wilderness, Pat soon learns a few tricks and spells which control the magical fabric of the universe ... and with the help of grimoires and perhaps a wise old mentor, struggles to learn more. By the middle of book three, this formerly inadequate stable clerk is a Mage who knows the True Names of things and can toss off Words of Power to command the forces of nature. It certainly beats working for a living.

(Greg Bear's novels *The Infinity Concerto* and *The Serpent Mage* offer one of the more intelligent recent versions of this rags-to-runes story.)

Now, the real-world version. Let's run through that familiar scenario again. Pat Nurd is a no-account but ever so aspiring young writer, who steps through the enchanted door of the computer shop and is pitchforked into a strange new world. Helpless at first in the menacing silicon wilderness, Pat soon learns a few tricks and commands which control the awesome fabric of the operating system ... and with the help of computer magazines and perhaps a local user group, struggles to learn more. By the time Pat should have finished book three of the current contract, this formerly inadequate hack is a Power User who knows three computer languages and can write programs to do all sorts of fun things.

Of course, Pat might not be doing so much actual writing....

If you accept the common picture of a writer as a mildly ineffectual person whose fantasies are acted out on paper, there's definitely a trap waiting here for the unwary. Behind the computer screen lies a strange and different space, a land which just like the average fantasy universe obeys new but suspiciously simple rules: rules that you can learn. All the sentences of its language are in the imperative – 'Do this, do that, for I command that it should be so' – and if they're phrased correctly they will always be obeyed. Gosh, what an ego trip. (Of course you have to watch it when invoking powers like the Demon Assembler – one misspelling, and this fiend can burst free of the pentacle to wreak havoc on your naked, vulnerable disk directory.)

We'd better not push this analogy too far. The general point is that, just like the standard untrained-wizard-makes-good plot, tinkering with computers can pamper one's little power fantasies. Issuing commands is fun. Getting a program to work is so much more definite and definable an achievement than bashing out another thousand words of prose which might well have to be rewritten anyway. I knew two married writers who were irrevocably split on the word-processing question: the wife wouldn't touch the husband's sinister devil-machine because, she reported, 'he keeps coming round and saying 'Hey, guess what!', and I say 'Did you finish the story?' and he says 'No, but I just figured out how to make the computer do something really great!"

All power corrupts, they say, and the absolute power we wield only in fantasies and programming ... corrupts absolutely, like Tolkien's Ring. You can fiddle round obsessively with a prime-number program or a deeply useless means of detecting split infinitives: you can feel you're doing things, achieving things, when the end result is a re-inventing of products rather less useful than a square wheel. Needless to say, this kind of skewed obsession with what's supposed to be a writing tool is not necessarily all that good for the writer.

Readers, this is no empty theorizing. Take pity on the neurotic, obsessed wreck which used to be a novelist called David Langford. The bytes have got to my brain cells. I've become a software company but it's a year or more since I actually wrote a book. Even if I can switch the creativity interface module back to writing mode, there is the ghastly fear that I'll end up producing a heroic fantasy trilogy, *Quest of the Silicon Mage*, in which the scorned hero discovers his ability to program the universe in Pascal, and ...

Semantics Corner

When is a piece of add-on software for LocoScript not a piece of LocoScript add-on software? The difference is that the first description is ideologically OK for anyone to use, while the second, according to Locomotive Software's eager and sharp-toothed lawyers, constitutes an attempt to defraud the public by implying that one's add-on program is written or approved by Locomotive themselves. It might sound daft, but be warned: lawyers have this magic power to see evil where others can't. I myself have utterly foresworn and abjured the marketing of LocoScript add-on software, and am nervously confining myself to the completely different field of add-on software for LocoScript.

8000 Plus 21, June 1988

The Leper's Squint

The great unsolved mystery of word processing (one of the most baffling and sinister cases, Watson, that I have encountered) is, why does it always look more convincing before you print it out? Somewhere between the screen and the paper, flawless-looking prose breaks out in horrid eruptions of typing errors, spelling mistakes and general infelicity. One interesting theory is that all printers contain devilish virus programs which not only pervert your text but transmit the changes back along the cable to your word processor's memory and disks. This idea, however, is obviously not paranoid enough to explain the reality.

But seriously ... here are some slightly more plausible thoughts.

Theory number two: even the superlatively wonderful Amstrad PCW screen, so much clearer than rotten old 256-colour ultra-high-resolution IBM Super VGA displays (he said with bitter irony), is not 100% easy on the eye. Even a mild visual fatigue that you don't actually notice can take its toll. Perhaps we're subliminally repelled, and tend to scan on-screen text less attentively than a piece of paper. (I know proofreaders who are *much* less attentive than a piece of paper, but this is a digression.)

Theory number three is deeply psychological. We do seem to put a ridiculous trust in computer infallibility. (When did you last check *all* the figures in a bank statement? In a recent fit of curiosity, Ansible Information found that although the visible arithmetic was OK, the bank had lost the entire contents of one account, put us in the red by debiting a cheque to the wrong name, and imposed a mysterious charge of several hundred pounds labelled CHEQUE BOOK. Take a bow, NatWest!) All those glamorously glowing, hi-tech letters on the screen look somehow convincing and inarguable and right, until they reach the paper....

Theory number four complements number three: this fast-lane world is so full of silly words, and stupid acronyms that look like misspellings, that those exposed to it become more tolerant of apparent mistakes in a computer context. Everyone who writes about programs sooner or later adds the non-word COM to their spelling-check dictionary, since otherwise every mention of PIP.COM or whatever will make the checker go tut-tut. This is fine until you mistype 'con' or leave off the end of 'come', and the spelling checker doesn't object. As you add more of computing's two- and three-letter jargon words and acronyms, there's an increasing chance that any short mistyping will correspond to a non-word and be passed as OK. (I spent ages wondering how the obscure formation RG had got into a spelling checker. It turned out to be part of my postcode.)

Theory number five is vaguely related to number two, and suggests that the problem is that of the Leper's Squint. This tasteful domestic feature was essentially a peephole in the thick wall of a mediaeval church or Great Hall, through which the afflicted were allowed to peer at the fun goings-on within ... a precursor of breakfast television. Reading a big document on *any* word processor has something of the same tunnel-vision quality. LocoScript aggravates this because the Leper's Squint is narrower than it need be – experienced users can't turn off the menu information and use the whole screen for text. The limited vista plus the slow shuffling of pages on and off the disk gives you more opportunity to forget the details of page 3 by the time you see page 6, and to use your favourite words 'concatenation', 'molybdenum' and 'gleet' too often, too close together.

How do you avoid or at least reduce the disappointment of finding your print-out less wonderful than the display's glowing prose? I hate to admit it, but there's no magic answer. Spelling checkers weed out many typing calamities, but have their limitations as above. 'Style checkers' also exist, but tend to concentrate on the red herring of complexity as measured by word and sentence lengths. (They also tend to be lazily programmed, so you have to muck around converting LocoScript text to ASCII files, etc.) Ultimately, you need to read the whole text through as slowly as possible, as often as possible.

A critical read-through at your normal speed is just part of the process. A second and *very slow* read is a frightful bore, but throws dodgy phrases into high relief; if you work alone, it helps to move your lips (shudder, shudder), subvocalize or even recite the whole thing loudly. There are plenty of phrases, especially in fictional conversation, which at first glance look all right but can't be read aloud without exposing flaws. In a recent bad SF novel, someone shouts, 'Sustained nullification on a huge scale might be beyond nature's ability to counteract!' The sentence is grammatical, just. But try shouting it.

W.H. Auden suggested the severest critical test of all, absolutely guaranteed to expose any stylistic flaw. The method is however only suitable for poets who aren't word-processor addicts, since it consists of writing out the whole thing again in longhand. Think about that, but not too hard.

More realistically, proofreaders sometimes look for misprints by reading in a way which kills logical sense and continuity, so that the mind doesn't treacherously make use of the context to correct lapses without even noticing them. I tried reading this very article in two such ways – backwards line by line, and backwards word by word – and can report that it's extremely boring. (Didn't even find any satanic messages.)

Style prose one's on effect bad a have itself could technique this using whether know don't I....

More On Silly Words

My favourite source of horrible acronyms and neologisms is the (otherwise nifty) trade newspaper *Computing*, which makes the whole thing twice as confusing with a house style whereby abbreviations are disguised as real words in lower case. The last word of the previous sentence, had it appeared just like that in *Computing*, would probably have meant Computer Aided Software Engineering – no, I'm not making this up. Similarly, 'eft' has little to do with newts (I think it's 'electronic financial transfer'); 'pcs' are computers; and 'risc', which you've probably met, is not a typo. I'm still waiting for them to refer to a dynamic RAM chip as a wee dram.

8000 Plus 22, July 1988

Another Useful Review

We were delighted to score an exclusive scoop by being the first magazine to review Grottysoft's long-promised SPUNG! SPUNG is an integrated development environment with built-in calculator, style checker, pop-up PacMan game, Serbo-Croat spreadsheet and IQ test. And once again, Grottysoft have a runaway winner on their hands!

Owing to production holdups, our demonstration PCW version of SPUNG was supplied on 24 disks for an IBM PC running in Amstrad simulation mode. Three special ROM packs and a co-processor borrowed from a BBC Master completed the easy-to-understand set-up, which will of course be even simpler when the final PCW program is available. For instance, SPUNG will be crammed on to a mere 15 PCW disks, losing only minor facilities such as the ability to process text files exceeding 150 words.

It takes only a few hours to instal SPUNG for your computer. The simple, on-screen, interactive installation program is a model of clarity: we knew we were in good hands as SPUNGINSTAL went straight to the important points by asking WAHT SORT OF COMUPTER IS THS?, DOES IT HAV DISK DRIVE(s)? and ARE YOU USIG MONITOR? By the way, it's necessary to run the SET24X80 program, load the SPUNGKEY.WP keyboard file and temporarily disconnect the printer cable before SPUNG will run ... we're sure these points will be mentioned in the production version of the manual, along with the need for a special power supply.

Amazing Innovations

It soon becomes obvious that SPUNG is a rich, powerful piece of fifth-generation software which can only grow in usefulness and indispensability with months or years of experience. The 25 minutes we were allowed with it (Grottysoft's Securicor man was on a tight schedule) did not perhaps exhaust the full range of its possibilities, but what we saw impressed us thoroughly. Pop-up, pull-down, multi-windowed, Y-front menus make the simplest functions available in as few as fifteen keystrokes – and there are short cuts too. The designers are not afraid to make imaginative use of the keyboard: the space bar, for example, has become an automatic 'toggle key' which moves you in mere seconds between the accountancy, graphics and desktop horoscope functions, and after a minute or two we found it much easier to enter spaces the SPUNG way, by holding down Alt, Extra and **f3**.

Error messages are signalled dramatically by reprogramming the 'beep' to sound like a large gong and flashing the screen on and off five times a second until you acknowledge your mistake (by pressing the Power button, of course – everything else is deactivated. We've never met a quicker way of teaching users not to make mistakes, especially since reloading SPUNG means fourteen disk changes).

That this is high-powered, fast-lane software is confirmed by the fact that it comes all ready set up to print out on Grottysoft's own-model laser printer, giving fabulous text quality for only £2200 extra. What's more, a Printer Configurator program is promised for next year, which will allow SPUNG's sensational graphics to be simulated on your old PCW printer too.

By the way, you'll need a memory upgrade, a hard disk, an add-on serial port, a mouse and a bar-code reader if you really want to make the most of the mindboggling facilities of SPUNG.

All This and More

The documentation is particularly impressive, more than 1000 pages of detailed and clearly-written information which should be translated from the Korean in no time. (Speaking of which, we're told that retail copies will offer a selection of spelling checkers besides the Rastafarian one supplied to us.) Despite all the challenges we threw at it, SPUNG always did very nearly exactly what we thought the manual said it should, and hardly crashed at all.

One warning for all you potential users: under certain rare circumstances, such as your pressing **f5**, Paste, Exit or Return, it is possible for the extra overhead of SPUNG's disk accesses to set fire to the floppy disk drive. We pointed this out to Grottysoft, and with their usual helpfulness they replied at once that nobody else had ever reported such a problem, that they would nevertheless write to their American parent company for advice, and that they could supply cheap CO_2 fire extinguishers as an added customer service. Our one small cavil has thus been effectively dealt with before the program's official release.

With such raw programming power and such in-depth support, it's remarkable that Grottysoft have been able to keep the price down to an attractive £199, exclusive of VAT, postage, packing and the special add-on software needed for SPUNG to accept text in lower case. No better bargain exists in the PCW market today, as the full-colour, multi-page ads in the next fourteen issues of this very magazine will testify!

Tailpiece

Although the tone of *8000 Plus* tends to be a mite more cynical than the above ... if you think that reviews reminiscent of this one never ever appear in any computer magazine, I can only admire the purity of your thoughts.

8000 Plus 23, August 1988

Best Foot Forward

One neglected literary form is the covering letter you send with your deathless manuscript. Thanks to industrial spies, I've secured several examples familiar to editors the world over. The challenge is to detect the subtle reason why in each case the recipient reached for his or her trusty rejection slip without finishing the covering note, let alone starting the manuscript. Match your wits against the professionals!

• Dear Editor,

What you're waiting for is a new idea to shake up the fuddy-duddy world of science fiction. Well here it is! Based on the mindbogglingly innovative concept of Earth being struck by a giant alien meteor with startling results, my novel *Lucifer's Footfall: The Forge of Earthdoom* is ...

• Dear Sir,

I see you publish BASIC programs, so you'll love my enclosed poem *The Joy of Babbage*, an epic in nineteen thousand heroic couplets. Mrs Gilbey of our village Literary Circle thought it was VERY INTERESTING and I know you will need no more recommendation ...

• Sir,

I cannot reveal my blockbuster plot to you as yet, since you would steal it and have it published under some house name by one of your tame hacks, thereby defrauding me of millions. I am on to the games of you 'publishers'. Before submitting the outline I want a firm contract guaranteeing a seven-figure advance and 110% of gross film rights. For the present I am not revealing my address – attempts to trace me and steal my notes will be useless. Kindly reply via the classified advertisement columns of ...

• Attention: Editor,

Revelations chap. xiii contains the clue. We know it takes Halley's Comet 76 years to complete one orbit but are you aware that if you add 2000 AD to Ussher's 4004 BC and divide the total by 76 it goes exactly 79 times? Since 1988 is actually the year 2000 this shows that the Second Coming will occur on 26 June. My manuscript conclusively proves ...

• Dear Mega-Ed,

I was having this totally ace game of *Bludgeons and Blackguards* with my friend Irving when we realized the excitement of our role-playing campaign would make an incredibly triff novel! So here, based on that month of fun, is *Lepermage of Elfspasm*, a brill fantasy dekalogy in which a lovable crew of Elves, Dwarves, Cats, Boggits, Men and a token Voluptuous Nymph go up against the Cold Dark Dread Force of Chaos Blood Death Evil, which ...

• Darling Editor,

I saw your picture in *The Bookseller* and at once knew we would become *very* close friends! I am 19 and *very* experienced. Perhaps we could have lunch together. Or breakfast. Of course I will be delighted to buy the meal! Don't you love champagne? Here is my photograph for you to keep. To fall in love sight unseen – it's like something from a mediaeval romance, isn't it? Speaking of which, I know you'd like a peep at the enclosed MS of my richly romantic historical novel, *I Was*

Edward II's Teenage Groupie ...

• Hi, Editorperson,

There's never been a novel like this! Imagine the excitement of a plot line in which all the past Doctor Whos meet up with Darth Vader, Superman, Gandalf, Marvin the Paranoid Android, Indiana Jones, Crocodile Dundee, Captain Kirk and Spock, Snoopy, Judge Dredd, James Bond, E.T., Mickey Mouse, Rambo and Cecil Parkinson! I am sure you'll have no trouble sorting out copyright problems, and then ...

• Deer Idiotr,

Plees find enclosed my novvle, it is handwrote Im afraid but you will not mind this becuase GENIEUS cant be mistakken can it? No retern post enclosed sinse this will nott be nessary as you will See ...

• Dear Sir or Madam,

The MS herewith is a *very first draft*. I could change almost anything on request. For example, in the slave bondage orgy scenes I am open to suggestions (your knowledge must be so much greater than mine). Just say the word and I'll alter the lard to cod liver oil, or the protagonist's name to – well, it's a teensy bit obvious, should we tone it down to Steelram or Goatfetish? Also there are details about bestiality which need checking in the light of your mature experience. I'm willing to take advice on any point. Just send a fully detailed letter of instruction and comment, and ...

• Editor, dear Editor,

Ever heard how George Orwell's best novels were bounced by several major publishers before they got to be international best-sellers? Well, history repeats itself, and my enclosed *Big Brother Farm* has actually been rejected by *exactly the same* wilfully blind publishing outfits as Orwell's. To add to the astonishing coincidence, I have chest trouble just like him. Knowing all this, can you *afford* to take the risk of not ...

• Dear Skiffy Editor,

This is a guaranteed SF best-seller – you don't even need to read it! My name will assure its success. I have the deed-pool documents all ready to fill in: the final decision is yours. Do you prefer Isaac Amizov, Alfred C. Clarke or Roberta Heinlein? I had also thought of H.G. Whelks, but do not think this would be such a good seller ...

• To Whom It May Concern:

Not merely a work of entertainment – my novel is *more*. Here in fictional guise are the truly shocking *facts* about the *conspiracy* of scientists, theologians and armed librarians who *control* us. Intentionally I have given over six chapters to *exposing* the jealously guarded truth about *gravity* alone – not a pull as *Communism* would have you believe, but a *push*! Unless you too are blind to reason or controlled by *laser signals* broadcast from *Chinese UFOs*, you cannot fail to ...

• Dear Gagged Lackey of the Thatcherite Junta,

Your lickspittle rag won't dare publish this, but ...

You see, of course, the common fault in all these? Not one of them addresses the editor correctly as, 'O Mighty Being From Whose Fundament The Illumination Of The World Proceeds'.

8000 Plus 24, September 1988

Bits and Pieces

Back in a period of such primordial antiquity that your present editor was but a gleam in some mad scientist's eye – that is, in 1986 – I wrote about the peculiar Amstrad spares situation. The trouble then was a clapped-out PCW keyboard, which according to Lasky's could only be replaced by purchasing a new PCW8256 system and throwing away the unwanted bits. Amstrad themselves, when quizzed on the subject, reassuringly quoted that favourite maxim of the computer industry: 'Go away, we don't do business with *end users*.'

(Perhaps it would be better if the big computer nabobs voiced their actual thoughts and called us all 'wallies' or 'suckers' instead. At least those terms sound as though they might apply to human beings.)

That was all long ago, and sources of replacement keyboards did emerge. One rather assumed that everything in the garden would by now be lovely.

However, just recently the trade newspaper *Computing* ran a piece headlined 'Amstrad users face long queue for spares', featuring such juicy quotations from the trade as, 'The system just doesn't work'. There are horror stories of six-month delays in replacing not merely old and tired keyboards from battle-scarred Amstrads, but dud parts from newish machines still under warranty.

One fascinating side issue involves the mighty clash of claims between a dealer who says Amstrad has been blaming hold-ups on a riot in one Taiwanese factory, and an Amstrad official who repudiates this vile accusation with the counterclaim that Amstrad can manage lots of supply problems without requiring any assistance from riots. But let's not go into that.

Spare parts were on my mind when I bought an Amstrad PPC to lug around on all too infrequent holidays. This represents a *lot* of brand-name loyalty, since the idea of unfolding such a hefty object on British Rail is a joke all by itself, without considering the expensive, short-lived, non-rechargeable batteries. (Use of the PPC's vaunted modem on battery power is prohibited with terrifying warnings – the general impression is that the mere attempt will cause the keyboard to melt and the disks to fly out like frisbees, while to your ears comes the horrid noise of Alan Sugar tearing up the guarantee.)

Actually this didn't bother me: I wanted to write in a holiday flat, not while sunbathing, bicycling or windsurfing. What would be nice, I thought, would be a spare mains power supply to save lugging that additional, bulky and not at all built-in adaptor box around. The dealer was tactful about my foolish request, and thought that Amstrad might make a few spares available in the mid-1990s, but for now – forget it, sunshine. There must be an opening here for an independent mains adaptor from some enterprising electrical firm? (13 volts, 1.9 amps, and do please send me a review copy.)

Where the spares situation becomes surreal is in the matter of the PPC's famous add-on colour monitor. Yes, there's a cute little monitor socket at the back, and by either software or hardware control you can switch from the LCD screen to the monitor of your choice ... but where to get one?

Various people have been patiently plugging in various IBM colour monitors, led on by naive trust in the fact that the PPC is nominally IBM-compatible. 100%

success is not reported.

It's said that standard NEC mono monitors work with the PPC. but with a recommended retail price around £200 they don't seem a terrifically attractive substitute for a colour version.

I borrowed a guaranteed all-purpose colour monitor from a friend, and plugged it in. All the monitor's fuses immediately blew and it hasn't worked since. My former friend is still digesting this information.

Dealers say, 'No chance, squire, *we* can't supply a monitor. Write to Amstrad.'

If you do write to caring Amstrad, they write back and say, 'Be advised Amstrad do not supply monitors separately.'

One anguished user responded to this as follows: 'It says in the handbook you can use the PPC with an Amstrad monitor. I only have a PPC. Can I buy a monitor?'

Amstrad: 'Be advised Amstrad do not supply monitors separately.'

User: 'How can I get an Amstrad monitor?'

Amstrad: 'Easy! Buy an Amstrad PC.'

User: 'Suppose I *have* a PC and my monitor has been stolen or caught fire; can I buy a replacement?'

Amstrad: 'Only if you send us the burnt-out bits or a statement from the police confirming the reporting of the theft.'

Blimey.

I'm not sure how this bodes for the oft-rumoured portable PCW. I hardly dare to hope that some enterprising hardware firm will start supplying cheap replicas of the remains of fire-ravaged Amstrad PC monitors. If so, please don't send me a review copy.

Tailpiece

On another subject altogether ... I take a morbid interest in the sufferings of the English language at the hands of computer programmers and journalists. Last month I read – in Another (Official) Magazine – a piece by Rex Last on German software. Its opening paragraph:

'One of the constant grumbles in the pages of the German computer magazines is that they have to put up with programs, manuals and adventures all written in English. It's a problem which, thank goodness, we on this side of the Channel don't have to face up to.'

How true, how very, very true.

8000 Plus 25, October 1988

Freelance Finances

Once, starving literary hopefuls would crouch in freezing garrets, scribbling masterworks by candlelight while rats gambolled underfoot. Nowadays garrets are hard to find (all converted to luxury yuppie apartments), starving authors all seem to own PCWs, workrooms mustn't get too freezing for fear of condensation in floppy disks, and probably rats are an endangered species. But the squalor of freelancing still has its charm ... though such writers spend long hours not writing but thinking about economics, and even longer hours wishing for some cash to be economical with.

Ursula Le Guin's advice to aspiring freelances was simply, 'Marry money.' Larry

Niven suggests getting your parents (like his) to put a million dollars in a trust fund for you. And austere James Blish warned against risking it until royalties from books written in your spare time exceed your 'real world' income (if any).

Q: OK, Langford, which method did you use?

A: Er, none of them really. As a sop to Le Guin my wife is at least solvent, and I heeded Blish by lining up two book contracts before I fled the Civil Service, but when I suggested Niven's method my father remarked, 'Pull the other one, son, it plays carillon chimes.'

A vital point when on your own is to *write everything down*. Yes, I'm sure that with an eye to future fiction you already jot down cruel word-pictures of people who sneeze glutinously into your face and tread on your toes in the bus. More usefully, hang on to bus tickets and every receipt for anything plausibly a writing expense, with a view to the coming tax return. Without tangible records you'll forget what you've spent.

Paying by credit card and treasuring the little greaseproof chit can be useful when (as with British Rail) getting a receipt involves surly reluctance and delay. But when reclaiming VAT, strict Customs & Excise inspectors won't allow any expense not backed up by a receipt carrying the supplier's VAT number ... so watch it.

Q: Where in my accounts do I put expenses for disks and printer ribbons?

A: Stick 'em both under Stationery.

Q: Can I claim the cost of my new PCW?

A: Eventually. However, a computer is that wonderful thing a 'capital asset', and to encourage massive industrial investment in new equipment the Government lets you claim only 25% 'depreciation' expenses each year. Pay £400 for a computer and you can allow £100 against profits the first year, £75 (i.e. 25% of the remaining £300) the second year, £56.25 the year after that....

Q: Blimey. You mean if I earn £400 and spend it on a computer solely for my writing business, I pay tax that year on £300 profit which I haven't got?

A: You're catching on. Actually, £300 total profit is a couple of thousand quid below the level at which you start paying tax.

My favourite cartoon shows this hooded character in loathsome rags, ringing a bell and calling out, *'Self employed! Self employed!'* Full-time writers tend to be self-employed, the exceptions being those who've set up limited companies to avoid graduated tax on an embarrassingly huge income. (Less wealthy authors trying this dodge find they merely pay embarrassingly huge sums to accountants, who probably suggested the idea for this reason.)

Self-employment gives you the privilege of paying Class II National Insurance contributions, which the DHSS extracts directly from your bank account to the tune of (currently) £4.05 a week, whether or not you're earning anything. Exercise for the student: program your PCW to calculate each month's cost, *bearing in mind* that DHSS months always have a whole number of weeks. *[2008: currently £2.20 weekly.]* Each year you're also done for Class IV contributions, a percentage of your taxable profit.

Q: What benefit does that bring?

A: None whatever. The Class IV rake-in is merely your governmental reward for becoming self-employed and forfeiting unemployment benefit. Sometimes the thrown-off shackles of former employment can look positively cosy.

Finally, the carefree joys of freelancing had better not be confined to writing. If your PCW muse leans mostly to poetry or short fiction, it's important to diversify. Even famous poets don't make a living from poetry: when not independently rich or mundanely employed, they live on editorial work, reviewing, journalism, reading for publishers, teaching, lecturing, media pontification, or writing articles in *8000 Plus* about how the PCW made it a doddle to produce *The Waste Land*, *The Faerie Queen* or *Beowulf*.

The list is similar for novelists, with one notable addition: hackwork. Ever wondered who writes those novelizations of obscure films ... that is, those not by Alan Dean Foster? Usually some temporarily broke author of moderate repute, who did a rush job of padding out a thin script for thick readers, and wisely used a pseudonym.

I've tried most of the above means of bridging the gaps between 'real' books. With practice they work addictively well, leaving no time for the Great Novel which you feel you should be writing....

Q: Oh come on, when I leave my job I'll have lots of spare time for everything.

A: It is a mysterious rule of freelancing that an entire day with nothing to do but write can somehow produce less than the few hours one used to manage in the evening after work.

Q: Well, why are you wittering on in *8000 Plus* when you could be writing chapter six of your sensitive comedy of manners *Sex Pirates of the Blood Asteroid*?

A: The money, chum, the money.

8000 Plus 26, November 1988

Misleading Cases

(In the High Court today Mr Justice Gleet summed up in the case of *Stupefying Software Ltd vs. Halibut*.)

Members of the jury, the facts of this case have already been put to you several times by counsel, with such matchless eloquence as to render them in all respects unintelligible. Numerous weighty documents have been placed in evidence, and although these purport to be elementary 'software manuals' devised for easy assimilation by the meanest intellect, it would perhaps not be wholly unjust to suspect that their meaning eludes you as it eludes me. Let me therefore strive to convey to you, probably for the first time, what this litigation is about.

Stupefying Software Ltd, as its managing director has informed this Court, is devoted to expanding the frontiers of knowledge, freeing mankind from mental drudgery, and (whether this be desirable or no) hurling its customers into the twenty-first century. To this laudable end, the company manufactures various useful computer 'programs'.

It is agreed that the defendant, Mr Alfred Halibut, purchased one such item, a light-hearted and diversionary game entitled *MegaRambo Nukefest*. Nor is it disputed, irrespective of the loathing with which one might regard it, that this was delivered in good working order.

Stupefying Software Ltd has consequently argued, with a smugness which you may or may not have found intolerable, that its part of the contract was amply fulfilled. Yet even the most bovine and comatose occupant of the jury box (I do not by this phrase wish to call undue attention to the snoring gentleman in the back row) must have dimly gathered that Mr Halibut disagrees.

The point at issue is an interesting and legally lucrative one, concerning as it does the unwritten aspects of the transaction. Let me strive to offer some examples sufficiently elementary for your limited comprehension. Were you, as a keen gardener, to order three tons of horse manure for the delectation of your roses, the technical fulfilment of this order would not impress you should the substance be unloaded in its entirety on top of your car. Were you the proprietor of a lodging-house whose regulations prohibited cats and dogs, you would not feel debarred from ousting a tenant who, while adhering to the actual letter of the law, had established in his room a small colony of wolves and a puma.

You may ask whether these analogies have any relevance, but I hope you will not, since should you do so I would instantly order your committal for contempt of court. Mr Halibut now claims that despite providing him with a superficially functional program, Stupefying Software has acted as unreasonably as the villains of my examples.

On receipt of his computer disk, Mr Halibut attempted to copy its contents for purposes of what is termed 'backup'. This process, the Court has been informed by authoritative if semi-literate expert witnesses, is to the computer user as important as life insurance, as psychologically vital as underclothing. You may therefore consider that on attempting to use his copy, Mr Halibut was rightly perturbed to be greeted with the message, 'STUPIFYING SOFTWARE THEIFGUARD PROTECTON SYSTEM, YOUR ATEMPT TO DO ILLEGAL COPYING HAS FALED HA HA!!!'

Despite the anguish and distress of hazarding a 'master' disk in actual use, the defendant was resolved to test his newly acquired educational product. This time he encountered the no less peremptory remark, 'THIS PRODUCT IS PERSONIZED WAHTS YOUR NAME ?' Having typed his reply, he was ejected from the program with the derisive retort, 'ILEGAL USER !!' By trial and error, and (as he has told this Court) the application of considerable intelligence, the defendant deduced the humiliating need to type his name as it appeared on Stupefying Software's receipt; that is, as 'A HALBIT', in capitals.

Mr Halibut admits that his *MegaRambo Nukefest* game thereafter functioned as advertised, displaying tasteful and graphically artistic nuclear detonations over the relevant tracts of South-East Asia. However, his pleasure was further muted by the fact that one-quarter of his computer's screen was effectively unused, instead showing the words, 'THIS POGROM REGISTERD FOR; A HALBIT 299 MAFEKING VILLAS LONDON NW27 UNAUTORIZED USE BY OTHER OR TRANSFER OF LICENSE IS ILEGAL UNDER COPYRIGT ACT PLEASE REPORT ILEGAL COPYING TO STUPIFYING SOFTWARE AT ONSE !!!'

The defendant claims to have felt deeply insulted.

Speaking for the plaintiff, the managing director of Stupefying Software has told the Court that without such basic precautions, Mr Halibut would be inevitably tempted to bulk-mail illicit copies to his numerous and unsavoury acquaintances, to advertise them for sale with criminally photocopied instructions, and to hawk them at less than cost on the streets of Singapore.

Here Mr Halibut's case strays from the broad paths of law and reason into the murky undergrowth of the controversial. Such prejudice and distrust, he alleges, left him thunderstruck. How, he movingly enquired until I was compelled to silence him, how could Stupefying Software imagine him capable of misconduct on this scale? His eyes having been opened to the corruption of the software world, Mr Halibut made haste to stop the cheque he had sent to Stupefying Software and which through an oversight had not yet been cleared.

For, as he argues and you may feel bound to agree, if such untrustworthiness is indeed prevalent, how could Mr Halibut be sure that forged copies of his cheque would not be disseminated to numerous and unsavoury computer dealers, or hawked at large discount on the streets of Singapore?

You may think this reasoning disingenuous. Repelled and nauseated though you must be by Stupefying Software Ltd and its products, you may feel that the company's action for non-payment is justified and must succeed. However –

(At this point the jury, all coincidentally computer owners who had struggled with ponderously protected disks, found Mr Halibut not guilty without leaving the box, gave him three cheers, and begged that all costs should be borne by Stupefying Software.)

8000 Plus 27, December 1988

Public Accolades

It's coming up to the time of year when some big mainframe computers get together and announce their intention of sending me wads of Government money. This is no empty promise: in February, for the sixth year in succession, I'll get a cheque for a few hundred quid. The payment and the amount don't even depend on my knowing dreadful and unsavoury secrets about cabinet ministers. In effect, it's democratically awarded by the great British public (you), and I shall be grateful all the way to the bank.

If you've ever published a book you are probably nodding your head wisely and murmuring 'Public Lending Right'. If you've published books but don't know this phrase, you're almost certainly missing out and should read on – likewise if you merely *hope* to get published one day.

PLR is a pleasant attempt at giving a fair deal to authors whose masterworks are borrowed a lot in UK libraries. Say you've written a novel retailing at £10 and get £1 royalty every time someone buys and reads a copy. (If you're a jaded, cynical author, you may not mind staying unread so long as the book gets bought....)

When the buyer is a public library, hordes of people will read the book, but the return to the author used to be the same: one 10% royalty fee only. Can't the author be rewarded a bit – that is, paid for public lending right as he or she might be for translation or film rights? For a long time there were arguments about charging a penny a loan and passing the revenue to authors: this would have been hellish to administrate in pre-computer days, and librarians were naturally dead against it.

The difficulties of introducing the PLR idea were compounded by the sort of conservatism which resisted other changes, like allowing women to vote. Authors have long been regarded as scum, and copyright protection is unique in law as being biodegradable: if you build a house it can be inherited by your descendants for decades or centuries, but if you write a book which still keeps on selling 50 years after your death, it goes into the public domain – anyone can reprint it, issue an edited version, etc.

How does PLR work? Now that so many libraries are computerized, the big PLR computer in Stockton-on-Tees merely has to gross up the borrowings recorded by twenty sample libraries: this produces an estimate of how many times each book by each PLR-registered author has been borrowed in the given year. I was cheered in 1988 to find that my own first book (published 1979) was reckoned to have had more than 3000 readers over the previous year.

Yes, all the authors in the scheme get statements detailing the statistics for every edition of every book: it makes for morbidly compulsive reading.

PLR funding comes from a not all that big lump of Government money somewhat arbitrarily allotted by the Minister for the Arts. After deducting running costs (for the computer and its acolytes), the remaining loot is divided up amongst the authors: the 1988 dividend was an exciting 1.12p per estimated loan. It would be less, but in order to weight PLR distribution towards starving as opposed to ultra-popular authors, no one is allowed to rake off more than £5000.

The PLR office is an oddly friendly department of the Civil Service: they actually answer queries in plain English and give unstuffy advice. What, I asked them, do I do about this book which I wrote with a pal who isn't eligible because he's a naturalized American? The reply: get him to waive the share of PLR which he can't claim anyway, apply for 100% of PLR income on the book, and slip your friend whatever percentage of this amount your conscience tells you.

Along with your PLR statement for the year, you get a fascinating leaflet of arcane statistics. Last time we were told that 57 authors had achieved the top whack of £5000, that 49% of those registered got less than £1000, and that the whole distribution was based on an estimated 639 million library loans grossed up from an actual sample of 7.6 million, only 235 million of the total loans being of books actually registered for PLR....

I need hardly mention that there are statistical checks designed to ensure that the estimated loans aren't boosted over-much by your entire family taking out all your books six times a week.

So how do you climb aboard? If you've published a book, and your name (or a pseudonym which is all your own) appears on the title page in company with not too many co-authors (books with four or more authors are ineligible on grounds of Too Much Work All Round), it's probably worth registering. I remember the anguished and envious groans on that very first PLR dividend day in 1984, from those who hadn't registered because 'obviously' the scheme would never bring them a penny. Oh yes – as well as all the above you must live in the UK and not be dead.

Registration itself is pretty simple; it used to be more tortuous, but the system has been fine-tuned since the early days. Basically, in the great tradition of our Civil Service, you fill in a form. All the information can be had from the Public Lending Right Office, Bayheath House, Prince Regent Street, Stockton-on-Tees, Cleveland, TS18 1DF.

Being a fairly obscure author myself, I'm still boggled by the fact that PLR brings me a little bonus each year, and also by the nuggets of accompanying information – like the fact that one book that never made me much in royalties is my most popular in libraries, with uncounted thousands of loans. I owe it all to you lot. And, of course, to the computers.

8000 Plus 28, January 1989

Down the Line

Your esteemed editor (whose name and face keep changing at inexplicable intervals) sometimes suggests that I write more about the seamy side of little software companies – that is, my own. Natural modesty permits me to do this only once a year.

Should you enter the rough world of software marketing, my main tip is: make sure someone else answers the telephone. Being slightly deaf, I have a permanent excuse. My Ansible Information co-director merely has a permanent twitch. Each morning he resolves to answer calls with utmost suavity and politesse; each evening he sends me a despondent report in which (I fear) the words 'bastard' and 'wally' figure prominently.

Before millions of readers rise up to lynch the entire Ansible staff and plough salt into the ruins, let me hastily add that most callers are just wonderful. As with street litter, grandstand violence and statutory rape, the problem is caused by a minority.

'What sort of minority?' you suspiciously ask. Aha.

• 'How do I know your rotten cheap software works? Can you send three evaluation copies?'

Translation: a rip-off artist. Who'd dare go into a bookshop, say 'How do I know this novel is any good?' – and demand to take it home and read it before deciding whether to pay? (Software and book copyright laws are the same.)

• 'Your brochure says your program will do this, that, and that. Will it really?'
'Yes.'
'Are you sure?'
'Yes.'
'Well, maybe. But will it advise me on diet horoscopes, do family trees for the gerbils and write to my VAT inspector?'
'No, that's not what –'
'Oh, it can't be much good then.' (Hangs up.)

Translation: here's someone who believed that stuff about how computers will transform your entire life, and is still looking for the one perfect piece of software which will instantly usher in the Earthly Paradise, at not more than £12.95 of course.

• 'You do this program for LocoScript on the Amstrad PCW, right?'
'Yes. It costs a mere –'
'Will it work with my home-made BASIC word processor on a Commodore 64?'

Translation: hope springs eternal in the human breast. Few C64s boast a three-inch disk drive.

• 'Hello! I'm having trouble with PIP, can I ask you nineteen detailed technical questions?'
'Look, sunshine, it's seven in the bloody morning and Ansible Information was in bed!'
'I have to get to work early! You should too! Now about PIP ...'

'Sorry, we're so overworked that we can't give our customers technical support for programs we didn't write ourselves.'

'Oh, I'm not one of your *customers* ...'

'Ritfim!' (Hangs up.)

Translation: the final expletive is constantly on support teams' lips, and is properly spelt RTFM, for 'Read the effing manual!'

• 'This is Megawally Associates Ltd. We're big, we're important, and we're taking no nonsense from you. I ordered your software weeks ago: nothing's come. This is urgent. If it's not on my desk first thing tomorrow morning, *there's going to be trouble!*'

Translation: a bastard. In ten cases out of nine, this call means, 'I told the purchasing people to send an order last week, and although they probably haven't done it yet, I'm going to take it out of you for not clairvoyantly realizing an order was coming.' Many frustrated executives find their egos soothed by this cheap, gratifying exercise.

• 'This is Megawally Associates Ltd. Our accounts department needs a receipted invoice for your software. Where is it? What are you going to do about it? I want it yesterday!'

'You'll find it in the parcel which at your loud and urgent request we rushed to you last Tuesday.'

'Oh, *that*. I haven't opened it yet.'

Translation: a right bastard. Sceptical readers are assured that both Megawally conversations are given almost verbatim.

• 'Send the stuff now and we'll pay in due course.'

'Sorry, we send on those terms only to educational, medical and government establishments.'

'Look, sunshine, we're one of the four biggest accountancy firms in the world!'

Translation: if so, they could scrape together a few quid from the petty cash, wouldn't you think? Despite attempts to be cuddly and nice, Ansible had to get tough about sending stuff on tick. 'One of the four biggest accountancy firms in the world' still owes us money two years after we were foolishly trusting, but generally the worst payers of all are computer companies. Amstrad themselves were forthright – when *they* demanded all our software, they made it clear that there'd be no nonsense about payment. Deep financial analysis of this proposed deal convinced us to save the postage.

• 'I'm going to have the law on you! You've destroyed my computer!'

'What!?' (Symptoms of heart attack, etc.)

'Yeah. I was running your program during a perfectly ordinary thunderstorm when the power-lines went down and in the dark I spilled coffee into the disk drive and trod on the keyboard, so it's all *your* fault. What are you going to do about it?'

'Er.'

Translation: we are going to take the phone off the hook for the rest of the day while we hide under the desk. Ansible Information seriously has been blamed for disasters resulting from loading/saving files while electrical storms raged overhead ... a good time to switch off and drink coffee in another room.

• (On the answering machine:) 'Please call me back to discuss your software. The number's 876543210. *(Pause.)* That's backwards, har har.'

Translation: some mothers do 'ave 'em. Only space prevents me from revealing much, much more. (Translation: *'That's enough whingeing'* – Ed.)

8000 Plus 29, February 1989

Writing More Gooder

I have been. Trying to. Improve my style and. Make it. Easier. For you punters to read. Is this any. Better?

Well, several recent articles and press releases have gone on about the virtues of style checking programs – software that reports on the clarity of your writing. My first paragraph is the result of paying rather too much attention to one of the most popular 'readability' tests. In future I will be writing lots more. Easy prose. Like this.

Conversely, would you care to measure your powers of analysis against a real brain-burster of difficult writing? A sentence whose comprehension, according to the best authorities, requires more than 31 years of full-time education? Fasten your straitjackets, check your MENSA credentials – here it comes!

'That excellent occasion of family celebration was enlivened by elephants, aspirin, carpentry and bananas.'

If that challenge was too hard for you, work up to it by easy stages via the following sentence of the same length – which the same formula calculates as being easier by some *25 years* of schooling.

'A sard pyx of lymph and gleet was limned with a quincunx of merkins.'

By contrast, what does the formula make of my first, disjointed paragraph? The answer which is cranked out is that it's suitable reading for a hitherto uneducated child after approximately one year in school.

Now for some explanations. The formula I'm using, or misusing, is Robert Gunning's venerable 'FOG index', the basis of most 'style difficulty' checkers. It dates from 1952; the name is an acronym, meaning Frequency Of Gobbledegook.

Working the actual FOG formula involves going through a piece of prose and making some simple counts which cry out to be computerized. You count the number and length of the sentences, and from that work out an average sentence length.

You also calculate the percentage of 'hard words' in the total word count, the rule of thumb being that a hard word is anything with three or more syllables which isn't a proper name (your program should be watching for capital letters), a combination of 'easy' words (this eliminates terms like 'horsepower' and 'superfluid'), or a variant of a word whose basic form is 'easy'. 'Edit' counts as easy, therefore 'editing' is easy. As a writer – no, no, an originator – I always suspected it.

Finally, add the average sentence length to the hard-word percentage and multiply this sum by 0.4. The result, Gunning being an American, is supposed to give the US school grade at which you should be able to tackle the prose. Grade 1 corresponds to six-year-olds, Grade 2 to seven-year-olds, etc. I'm uneasy about adding two different kinds of number (an amount and a percentage) at the last step, but it seems that the formula tests out pretty well on a practical basis.

How ludicrously it can fall down is shown by those slightly spurious examples. My opening paragraph has no 'hard' words at all and fiddles the sentence-length average by simple defiance of grammar. The 'excellent occasion' sentence rates as terrifically obscure since 64% of its words are trisyllables and thus 'hard', even if they don't look hard to you. And the 'sard pyx' fools the rule-of-thumb difficulty gauge with a cluster of short, 'easy' obscurities.

Obviously the second and third cock-ups can be partly eliminated by a program with access to some comprehensive dictionary of genuinely easy words. The first is actually more subtle. Breaking up sentences automatically wangles you a better readability score even when the. Disruption of logical. Flow means. A harder. Read ...

It is time for a little Viewing With Alarm. This hasn't merely been an exercise in poking fun. Style checker programs are loose in the world. I don't know whether any author who submits text on disk has achieved the dubious privilege of being the first to be rejected unread because 'the style program says you're too highbrow for our median market'. It will come.

Fortunately the checking contains the seeds of its own ruin. I once wrote a spoof fairy tale in which a king, well read in the literature and having to choose between the merits of three princely suitors for his daughter's hand, decides to cut out the complexities and go straight to the inevitable winner by simply asking which prince is the *youngest*. Later it occurs to him that this point is too well known, and all but the youngest prince will infallibly lie about their age.

Just so, authors may lie about their style, as in my examples above. Gunning's FOG formula assumes that writers are unspoilt and don't contort their prose to achieve a 'correct' score. Authors with style checkers of their own may soon be cunningly aiming for the exact readability level demanded for the chosen market.

But by this stage, what will 'readability' mean? Only that the prose is tailored to get the right score from tests like the FOG index. Some editor is still going to have to toil through the stuff and find out whether, like my second example but unlike my first and third, it really is readable. Back to square one, everyone.

Meanwhile, an outfit called Scandinavian PC Systems is flogging a style checker called READABILITY, presumably not for the PCW market since here you can't get away with a price tag of £59.80. Is it any good? A rave review appeared in *PC* magazine, excitingly illustrated with screen prints of the program at work. As it improves your English, READABILITY can be seen to offer you *choises* (sic), to give an overall *evaulation* (sic), and to report how many short or long words you use *in average*.

Truly, this could be the software that teached we to write good!

8000 Plus 30, March 1989

100 Years Ago

Welcome, one and all, to the second monthly number of this family periodical. Your warm response to our first has gratified the Editors most excessively. No doubt of it: scattered through the length and breadth of England are thousands of happy devotees of the machine from which our publication takes its name – none of whom would willingly return to the old, wearisome ways, yet most of whom are a little 'stumped' by the mechanical intricacies. It is for you that we publish *Remington Plus*.

Without further ado we present our first 'Enquire Within Upon Everything' page. Herein readers' queries will be answered monthly by 'Aunt Davinia', who seeks to be to you a Counsellor, Guardian, Instructor, Companion and Friend.

• 'Baffled' – We agree that the pamphlet of instructions provided with your machine is shamefully deficient. The reason that your letters are difficult to read lies in a point which was not made clear: you are meant to insert a sheet of paper into the typewriter by dint of rotating the knurled knob at the top right. If you do not follow this somewhat technical explanation, you must consult the tradesman who supplied your machine. You will find that the use of paper makes the result far more legible and also less costly to post. (Should you wish us to return the roller carrying your letter, you must remit eighteenpence for carriage.)

• 'Paterfamilias' – An excellent spelling checker is published by Bowdler & Company, containing none of the terms to which you discreetly allude. Your children may consult it without fear. However, such lexicons do not offer the alternative renditions you require. The 'add-on software' of M. Roget provides this facility but includes notions somewhat too sensational for those of tender years.

• 'lowercase' – This is a frequent enquiry. Although the feature for which you ask was in fact introduced in 1878, it has never in our opinion been adequately documented. Somewhere amongst the keys of your machine – we cannot say where without further information – you should find one bearing the legend *Shift*. Depress this firmly when typing the first letters of proper names and the opening words of sentences. With a few weeks' practice you will soon master the trick of releasing this key before the next letter is typed. Space precludes our entering into fuller details, but for our summer number we are preparing a lengthy essay entitled *Shift Lock Secrets*.

• 'Pro Bono Publico' – Clearly you have been dulling your mind with the childish fantasies of such as M. Jules Verne, until your sense of proportion is quite eroded. The notion that personal typewriters might be driven by some such fantastic means as electricity is sheer nonsense. Only the huge 'mainframe' typewriters of commerce, such as that devised by Mr Edison in 1872, could conceivably operate in such a fashion. Are you afraid of honest toil?

• 'Thrifty' – You should read our pages with greater attention. Each issue contains several advertisements for preparations of soot, lamp-black, boot-polish, *et cetera*, supposedly suitable for the 'rejuvenation' of ribbons. We cannot recommend any one in particular. Your own efforts with oak-galls and household chemicals are perhaps commendable, but imperfect: the ink emits noxious fumes which made our chambers uninhabitable until your letter was removed by the public hangman. We have forwarded his account to your address.

• 'Aspirans' – Certainly not! Under no circumstances should any English author take colonials like Mr Mark Twain as setters of precedent in this respect. We dare say publishers in those uncouth parts are prepared to accept typewritten submissions. In Britain, literary etiquette is unchanged. You may prepare drafts of your work on the machine, but must write it out in 'letter quality' script for the publisher. Vellum and quill pens remain optional.

• 'Fun-Lover' – Yes. Although the Remington is almost exclusively thought of as a serious machine for the use of authors and businesses alone, many games are indeed available. For example, the advertisement from 'Diversions and Pastimes Workshop' on page 94 offers a selection which includes 'Consequences', 'Acrostics',

'Double Acrostics' and 'Postman's Knock'. Many enthusiasts declare the Remington adaptations of these games to be infinitely superior to the old 'manual' versions.

• 'In Statu Pupillari' – Your tutor or governess has sadly neglected your education. A child of your years should be aware that the typewriter, far from being a new idea, appears in a British Patent as long as ago as 1714. Nor was your letter free from errors of grammar, syntax, layout, punctuation, spelling and diction.

• 'Blackbeard' – We can take no pleasure in your account of how (no matter with what ingenuity) you have 'hacked' the internal mechanisms of Remingtons to effect various claimed improvements. Such tampering is in clear violation of the makers' warranty conditions. We hold it our duty to prevent such bad and dangerous suggestions from reaching our readers. After taking legal advice we must also return your article on the use of a modified machine to generate 'official' identification and thus penetrate the Bank of England's ledger system. The notion is in poor taste.

• 'Wordsallruntogether' – The key you seek is the very wide one which you will find lies closest to you as you operate the mechanism. Contrary to your somewhat petulant implication, we consider the manufacturers to have labelled this clearly and correctly with a picture of a space.

Aunt Davinia will return with further enlightenment next month.

8000 Plus 31, April 1989

Nebulous Statistics

Wearing other hats in other magazines, I also review science fiction ... but I never thought I'd cover an SF book here. It's not so much the book as its introduction, which unites two favourite themes: the unreliability of statistics and the often hilarious results of applying machine analysis to our slippery English language.

For an enjoyable if dated look at the first theme, see Darrell Huff's *How To Lie With Statistics* (1954). This covers many classic cock-ups, like the American opinion poll which selected its thousands of victims totally at random from the telephone directory but was still dead wrong about the election. (Phones weren't quite as universal then, and the selection method ruled out hordes of less well-off voters.)

As for my jaundiced views on how computers look at words ... see the March 1989 column on page 65.

Here's the book, *The Best of the Nebulas* edited by Ben Bova. The Nebula awards for SF stories are voted on by members of the SF Writers of America, an organization which anyone can join on the strength of three short published stories. They've been presenting these awards since 1965; this is a 'best of the best' anthology.

What's perched by my keyboard is an uncorrected proof copy from Tor Books (New York) ... so before actual publication, some editor might yet remove the points which caused me to say loudly, 'Oi!'

As early as his third introductory sentence, editor Bova gets into trouble. As judged by SFWA, the anthology 'contains absolutely the best SF stories published between 1965 and 1985'.

Oi! Here's a thought experiment. Imagine that 1980 was a terrifically good year and saw the publication of the best, second-best and third-best short SF stories of 1965-85. Only one could win the award. (We'll assume it's the best, although short-term log-rolling is rife, and as Bova admits, opinions shift with time.) Then the SFWA members who voted on the contents of this anthology couldn't possibly pick the second- or third-best stories: they won no Nebula and are excluded, unlike the fourth- and twentieth-best, which *did* win in leaner years.

Then, how to conduct the actual voting? Bova polled North American SFWA members only, since: 'Overseas membership is too small to have a significant effect ...'

Oi! Two more thought experiments. One: our government excludes all 'fringe' parties like the Green Party or SDLDDLSP from the next election because they're too small to affect results. Two: Bova notices that SFWA has only one member in, say, Kansas. Does he exclude Kansas on the same logical ground? Does he hell.

(One reason SFWA overseas recruitment remains small is that members in funny foreign countries like Britain feel they're not considered 'real'. Bova certainly adds to this impression. Example from my own experience when a member: SFWA refused to handle a dispute with my publishers, which is supposed to be one of its functions. The UK Society of Authors browbeat said publishers into a cash settlement for crimes in breach of contract. Moral: join the Society, not SFWA.)

Best of all is the Bova description of how all the voting questionnaires were analysed by fantastically sophisticated software which actually ... compared the frequencies of the words used!

This sounds suspiciously like the program I wrote years ago and called GREASE in honour of David Lodge's nifty novel *Small World* (where similar software plays, as it were, a bit part). I meant it as interesting and maybe illuminating fun; Bova says it's a powerful tool for determining covert opinions and attitudes.

Thus, he points out triumphantly, the words 'fiction' and 'science' turned up lots of times and were about equally frequent, presumably betraying the 'not outwardly expressed' fact that this poll concerned SF. Warming to his theme, he notes that the most frequently used words of all are: 'story' and 'stories'.

The point soon emerges as Bova launches an attack on the artsy-fartsy poseurs who filled in voting forms. These fibbers, he complains, kept banging on about 'literary quality', 'passing the test of time', and 'impact on the field', all esoteric virtues of which Bova's own SF is certainly innocent. But the computer saw through this holier-than-thou posing!

For, as Bova reports gleefully, the most frequently used words of all were 'good' and 'read'. At heart the voters spurned the high-flown litcrit stuff and wanted a *good read*....

Oi!

Here are some questions to ponder.

• How come the most frequently used words were also 'story' and 'stories'?

• Might the frequent words 'good' and 'read' have been parallelled by frequent use of 'not', 'just' and 'a'?

• Might 'read' have sometimes been a verb?

• Aren't there many more ways of phrasing subtler literary virtues than there are of saying 'a good read'? For example, mightn't you talk about sheer writing *quality* or *excellence*, and wouldn't the analysis be meaningless – or even more meaningless – unless the word-counts for 'quality' and 'excellence' were combined?

• Are you now surprised to learn that in pursuance of some murky sub-argument, Bova proudly points out that 'quality' scored high while 'excellence' came low?

• Would you buy a used statistic from Ben Bova?

Probably his conclusions have several grains of truth; but conclusions reached by such shabby reasoning are automatically devalued.

Meanwhile, you're lucky this isn't an SF magazine, otherwise now I'd have to go on and review all the collection's actual stories....

8000 Plus 32, May 1989

• *Suffice it to say that for lack of room to reprint the novel winners, Ben Bova instead offered little essays on those books; and that to remove all possible bias and ensure trenchantly impartial evaluations, these essays were written by the book's authors.*

The Long Goodbye

I sat in my office watching the level of the bourbon slowly sink, like Amstrad profits. Another couple of slugs and I'd start kidding myself I could understand CP/M BDOS error messages. Being a private computer consultant can drive you crazy.

The client was a dame. A dame whose picture would melt anyone's video digitizer. There were lines on her face, maybe thirty-two of them, but she'd stayed easy enough on the eye. She'd said her name was Joyce.

'I need you to track down some disks.'

'LP, compact, brake or spinal?' I quipped.

'CF2,' she snapped. 'I need them bad. My supplier says there's a shortage....'

'Of all the columns in this magazine,' I sighed, 'you had to walk into mine.'

So now it was my problem. Maybe this tied in with my April investigation. An outfit called M.D. Office Supplies had put a flashy ad in *Amstrad PCW* magazine. It offered disks. Five and a quarter inch disks. Three and a half inch disks. No other size at all.

Dealers ... who can figure 'em?

I started out with my own dealer contact. He lived in a bad part of town, where the ads are small and cheap. Down every dark alley someone was waiting to clobber you good with VAT and carriage, not included in the advertised price. In that district, customer support meant that instead of throwing you out they carried you out. And dropped you.

'Times are hard all over,' muttered the dealer, nervously shifting a wad of greenbacks from one pocket to another. 'Our wholesaler, he's got a million disks back-ordered. Nothing I can do to speed them up, squire.'

I got the grimy sheet of paper from my pocket. 'These are the discount prices you promised you'd hold level until ...'

I'd forgotten his assistant. I spun around, one hand diving for the holster, but too late. He smiled as he slugged me behind the ear with a fat list of revised prices. My Visa card went limp like a Dali watch. On the second blow the whole world curled up and turned black as NLQ from a fresh new ribbon.

The last words I heard were, 'Thirty days minimum delivery time.'

... My brain kept throbbing and reporting 'missing address mark in frontal lobe'. My mouth tasted like the toilet floor at a computer show. It was no time to be talking to the client. I was talking to the client.

'How about some results?' she said with a green light in her eyes.

I explained: 'Down these mean disk manager menus a man must go who is not himself mean, who is neither tarnished nor afraid.'

'I guess that means no progress.'

'You got it.' I patted her on the monitor, which was curved in all the right places. 'Here's looking at you, kid.'

Following a hunch I checked out another lead, another of the million outlets in the naked city. These were big-time mobsters with an open plan dump full of potted plants. Out of my league. One false move, and the heavy in the smart suit and tie would have me up against the wall. I'd be lucky to get out without being sold an 80386 IBM system with VGA display and laser printer.

'I want to know about disks. CF2 disks.'

His laugh would have made a hyena hide nervously under the bed. 'Get out of here, you small-time punk.'

I got mad. He froze as I drew a bead on him with my snub-nosed RS-232 interface. 'You've got contacts,' I told him. 'Guys who spend fifty big ones on accounts programs, and that's just to handle petty cash. Guys who know the disk situation.'

'So?'

I fired seven data bits over his shoulder, to show I meant business.

'You're thinking to yourself, is his RS-232 set up for seven-bit or eight-bit transmission? You're wondering, has he got one bit left? What you should be asking yourself is ... do I feel lucky?'

His teeth did castanet impressions. 'OK, I'll spill it. It's the import connection – most big Jap outfits have pulled out of CF2s. Only Maxell still make them. There's a famine. The operators with big stocks plan to clean up.'

It fitted together. I dived through the glass doors just before a fusillade of hype could shoot fatal holes in my sales resistance. This is a lousy stinking business, I thought as the pavement came up to hit me.

Later: 'I can't touch the Mr Big behind this,' I told the client. 'It's the old story – he's out of reach. Seems he made a killing installing cheap drives in a million machines, and left the suckers to feed them with expensive disks. It's easy to get hooked ... hard to kick the CF2 habit.'

'I can pay,' she whispered.

'Seems the operation never went over big, Stateside. Maybe the Mafia didn't like the interference, maybe the punters saw through the scam. Without that market, the Japs got leery and dropped out.'

She blazed greenly at me. 'So you can't fill my order. How d'you plan to stop me blowing the whistle on this sleazy consultancy business of yours?'

There's no arguing with dames. 'Farewell, my lovely, until the shortage is over....'

My fist caught her smack on the power button, and she slept the big sleep.

8000 Plus 33, June 1989

Fizz! Buzz!

What ghastly, warping childhood experiences could make someone grow up to be a freelance futurologist and software know-all? Revelation came to me at a computer show where I'd arranged a second mortgage to buy a drink. It tasted awful, and I thought of when a pint almost as bad cost me exactly one-twentieth as much.

This thrifty fluid was some foul fizz served in South Wales pothouses to schoolboys of limited taste. My teenage memories started trickling back. Proust sailed into the wastes of lost time at the remembered nibble of a biscuit, but I was made of sterner stuff. The remembered tang of iron filings ...

Participants in those smoky pub sessions were thrown together by friendship, throbbing absence of girlfriends, and the natural human urge not to be doing homework. It was my evil pal Dai who enlivened the evenings with the direly hazardous game Fizz-Buzz.

If you're lucky, you won't have met it. Semi-drunken clowns sit in a circle, counting aloud, clockwise round the ring: 'One.' 'Two.' 'Three.' At five, and every multiple of five, the current sucker must instead cry 'Fizz!' At seven and its multiples, the word is 'Buzz!' and the order of play reverses direction. Anyone failing to make the right noise at the right time must take a huge swig of beer (amateur rules), drain the glass and buy another (tournament rules), or knock back all visible drinks and buy a round (insane idiot rules).

Well, it beat South Wales's other conversational topics: women (frustrating since none of us knew any) and rugby – even more frustrating since, precociously beer-raddled, we couldn't play the national game without wheezing and falling over. This has been a health warning.

There was a weird satisfaction in doing this daft business right, 'the solemn intoxication which comes of intricate ritual faultlessly performed' (thus Dorothy Sayers on bell-ringing) – except that the ritual wasn't *that* intricate. Even the double thrill of 'Fizz Buzz!' at multiples of 35 failed to reach orgasm level.

Clearly the 'game' lacked intellectual challenge, at least until so late in the evening that remembering one's name also began to present difficulties on the order of Fermat's Last Theorem. We tried attaching electrodes to the sluggish rules. An early experiment was to assign 'Oink!' as the, er, buzzword for multiples of 3. Dai soon developed a particularly obscene 'Oink!' whose mere enunciation counted as gamesmanship. The corpse of the rotten game began to twitch.

'Burp!' for multiples of 11 was the next logical addition. By now, some us were sweating, concentrating intently, and falling over sooner than of yore (see above: Tournament Rules). Then came a quantum leap into genuine mathematical abstraction: 'Clang!' each time the count reached a prime number. (After savage debate, the dogma of mathematics was cast aside and 1 was declared prime.) Around then I stopped remembering petty things like closing times or how I'd got home afterwards. Sanity finally died with the two-pronged introduction of 'Pow!' for perfect squares and 'Zap!' for powers of two. Was 1 a perfect square? (Oh, all right.) A power of two?

By now, alert readers will see, there were no blasted landmarks. Pale, strained faces ringed the table, struggling to follow a count which began not 1-2-3-4 but: 'Clang Pow!' 'Clang Zap!' 'Oink Clang!' 'Pow Zap!' It was a supreme moment of triumph if we successfully galloped into the straight with 'Oink Buzz!' 'Burp!' 'Clang!' 'Oink!' 'Fizz Pow!' ...and, at last, the first number in our counting system which came through in clear.

'Twenty-six!'

I've never worked out what the pub regulars thought of us, but they used to look worried.

The suggestion of 'Ping!' for cubes was perhaps unnecessary. Perfect numbers also received short shrift. The sessions ended after a serious plan to signal numbers in the Fibonacci series (1-1-2-3-5-8-13 ...) with, appropriately, 'Argh!' Rather than debate whether 1 should now be intricately coded as 'Argh Argh Clang Pow!' owing to its double appearance in the series, we all went to university instead.

In Oxford, many splendours and miseries followed, but the demented game wasn't so easily escaped – not merely because I inflicted it on precocious hackers who programmed the Nuclear Physics Department computers to list every response up to ten thousand. (When I write *Advanced Fizz-Buzz – the Dungeon Master's Guide*, I'll know where to do the research.) Those nonsense sequences were strangely hard to shake off. People have been driven round the twist by obsession with Charles Hinton's coloured cubes for visualizing the fourth dimension (1904). Not being quite intellectually up to that, I still suffered years of fizzes and oinks and clangs running round my head like mathematically-minded squirrels.

(Also I invented variants like Real Men's Fizz-Buzz, played with all the real numbers between 0 and 1 with special grunts for transcendentals – *you* go first, thanks; Big Fizz-Buzz, in which anyone reaching the first transfinite ordinal before closing time must intone 'Someone's Been Cheating!'; and, after a crippling attack of Douglas Hofstadter ... Self-Referential Fizz-Buzz incorporating Strange Loops.)

The sound effects in my skull did eventually fade, but as a possible side-effect I seem to have spent my working life doing vaguely mathematical and computerish things, from SF to doomsday-weapon simulations to making PCWs count words. Wasting time in pubs? I can almost truthfully say: 'I owe my whole career to lousy bitter and Fizz-Buzz.' Death comes on swift wings to anyone who responds, 'What career?'

Admittedly, my failure to get rich by writing *The Hitch-Hiker's Guide to the Galaxy* can be blamed on schoolday conditioning to think that, for the reasons above, 26 is an infinitely funnier punchline than 'Oink Buzz!'. I mean, funnier than 42....

8000 Plus 34, July 1989

Paperless Publishing

Each year, as technology marches resistlessly onward and hordes more writers discover the joys of automation and repetitive strain injury, the image of word processing grows friendlier – more commonplace. Even non-technophiles can play; even those organizations most mired in ancient tradition, the book publishers. I once moaned here about laggard publishers who despite your offer of perfect copy on disk would still insist on having the whole thing retypoed at enormous expense by their palsied printers. This still happens, but nothing like as universally.

(However, do note that preparing clean print-outs of your stuff remains important. A publishing outfit which favours the IBM GrottyScript word processor will not be happy to get an unsolicited novel on a three-inch disk. These days I add 'Available on disk in such-and-such formats' on MS covering sheets.)

The point of disk submissions is only partly that publishers save money on 'rekeying'. Authors may or may not get a share of the resulting extra profits ... but for a dedicated author, the great attraction is that your words should be printed as you wrote them.

Of course, merely changing the format of your text for book pages or magazine columns might not be the end of it. Discreet corrections could be needed, where the sensitive, intelligent editor realizes that you've cocked it up.

In addition, there's the question of house style. This makes sense in two contexts. First, if an author is just plain inconsistent about whether to write 'World War 2' or 'World War II' – 'realise' or 'realize' – 'no one' or 'no-one' – then the publishers will quite reasonably get out their 'style book' and standardize on their own preferred usage. Second, magazines routinely do the same to articles for the sake of coherent presentation.

So far, so good ... but horror stories follow. The trouble arises when publishers mechanically impose their wretched style-book on prose which is consistent and correct, but doesn't conform to the 'rules' scribbled by some bored editor in 1966. For example, at school we all learned to use double quotation marks for speech: "Hello" and not 'Hello'. Many publishers go through entire books substituting single quotes; on investigation you find that half the editors you ask would prefer double ones, but 'we have to change it because it's the house style'. *[In fact I yielded to single quotes simply because – this was before smart quotes became standard – proper left-and-right ones were easier to enter and looked better on the computer than the double equivalents. Yes: blame the machine.]*

That's a minor annoyance. One outfit drove me to distraction by preferring the 'ize' forms of verbs (OK by me) and as a result changing every use of 'advertise' to 'advertize' (wrong by either standard), and 'laser' to 'lazer'. The same publisher refused to print the book's dedication because it would injure their lexical dignity: 'To XXXX, who teached I to write good.'

Another – and let's name names, it was *Practical Computing* in the days when they ran occasional fiction – favoured me with the worst copy-editing job I've suffered in my life. It was a humorous story, and they passed it to a dour technical editor who carefully removed all the funny bits and rewrote the horrid abbreviations which his house style forbade. Thus I'd have someone saying 'Can't fool you!' and it would come out as 'Unable to fool you.'

But then I started submitting (by arrangement) on disk, and of course everything was perfect. Er, well, um. What is actually amazing is the discovery of how you can send perfect prose on disk and errors will be *put in* by eager copyeditors. I still wake up screaming at the memory of my article on fantasy for a magazine whose editor couldn't spell Tolkien, and who carefully uncorrected each of my eighteen mentions of the guy.

Then there's the magazine for which I write monthly book reviews. Sometimes I dream of sneaking in and finding out what they *do* to the disk file to lose random letters here and there, just as though it had been retyped by hand. I suspect technology can't be blamed for their dismemberment of the unfortunate US editor and anthologist Beth Meacham, who was finally printed as Beth Full Stop New Paragraph Meacham.

A weekly computer rag not a million miles from Bath suffered a severe lapse of its spellynge chequer when I sent (on IBM disk) an article which frequently mentioned the word 'twilight'. This came out, consistently, as 'twighlight'.

After all this bitchiness, you're doubtless expecting me to finish by belabouring *8000 Plus*. Unfortunately for lovers of bloodshed, they're pretty good, apart from a tendency since issue 30 to lose my italic marks. (*Do shut up about that, you whingeing sod – Ed.*)

True Confession time: I must admit that when recently confronted with a novel safely stored on disk, I didn't do too brilliantly myself. This awesome masterwork was my and John Grant's horror-novel spoof *Guts!*, of which it has been said, by my wife, 'Yuk!' Mr Grant and I had drafted it on word processors; alas, he'd used his 8256 while I'd been trying out WordPerfect on an IBM clone. (There were complicated reasons for this. No hate mail, please.)

We thus had a book whose chapters were in different formats on different-sized disks. Fortunately, I'd been tinkering with a program to convert LocoScript files to other formats while avoiding the ASCII route which loses all the italic/bold/underline markers. *Guts!* became a huge and revolting guinea pig for this software.

After the revisions were made and the book was printed out, I found a slight problem whereby (for reasons which almost made sense) every dash in the novel had vanished in transit. During my long afternoon of checking early drafts and inserting all the lost dashes by hand, I was mortified to think of you lot chuckling, 'Ho ho, serves him right for straying from the one true way of LocoScript....'

8000 Plus 35, August 1989

Little Dots and Squiggles

I've been groaning my way through more unpublishable typescripts, and suspect it's time for some tub-thumping fundamentalism. What passes for punctuation in these benighted days is quite frequently enraging. Sage advice and maddening pedantry follow herewith.

Apostrophes. If you write 'it's' as a possessive pronoun, editors will call you illiterate. (Its only correct use is as a contraction of the well-known phrase or saying 'it is'.) Beware of Grocer's English, where the apostrophe is used for all plurals: 'tomato's' instead of 'tomatoes' and so on. Many people get confused by possessive plurals and words ending in S: the pips of several tomatoes are 'the tomatoes' pips', but Steve Whatsisname is '*8000 Plus*'s editor', not '*8000 Plus*' editor'.

Brackets. I use too many ... do as I say, not as I do. When writing English as opposed to mathematics, resist the temptation to flaunt the PCW's square, curly or angle brackets. (However, if you ask nicely I'll permit you to use square brackets to distinguish a parenthesis within a parenthesis [like this].)

Colons. The colon is tricky because it has two uses: introducing a list (as here) or example, and, more rarely, linking two sentences to point up their contrast. 'I am a columnist: you are not.' Business English tends to put a superfluous dash after a colon which introduces a list – but let's stick to English English. ('Who is this guy Colin Dash?' said my American pal.) Many Americans capitalize the word following a colon. This is incorrect, even according to many other Americans, but is spreading over here thanks to cheapskate publishers who photo-offset from US books.

Commas. These are most often misused as an illiterate means of stringing sentences together, for example this 'sentence' should be broken into two with a full stop or given another punctuation mark instead of its comma. (SF author Harry Harrison is a persistent offender in this respect.) Warning to 8256/8512 owners: as your ribbon fades, keep an eye on the tails of printed-out commas. They're the first things to vanish when greyish print is xeroxed, and prose doesn't half look illiterate when all the commas turn into full stops.

Dashes. Thank goodness, we've escaped the elegant anonymity of past centuries' dash-spattered novels: 'In the year 18 – a young man might have been observed purchasing a copy of *8000 P* – in the town of B – . He glanced within and ejaculated, 'D – !" The dash is a more frenetic and breathless version of the colon, which can also mark parenthetical phrases like ersatz brackets or commas. How to type it? Space-hyphen-space is common, but sometimes this slips into print as a mere hyphenation. Space-hyphen-hyphen-space makes your intention clearer. Some writers prefer double or even triple hyphens with no spacing at all.

Ellipses. See full stops ...

Exclamation marks. Use them very sparingly!! There's no grammatical rule against slapping exclamation marks on every sentence you think is dramatic, clever or witty! However, this is the literary equivalent of laughing loudly at your own jokes while digging violently at the listener's ribs!

Full stops. You must have noticed them, those little dots at the ends of sentences. Put three together and you have an ellipsis ... like that. Many publishers like you to put a space before three dots. When ending a sentence with an ellipsis, pedantic writers use four dots.... Don't overdo this: it's a way of nudging the reader to hint that Things Are Being Left Unsaid, and (as with exclamation marks) people resent too much nudging.

Inverted commas. See 'quotation marks'.

Parentheses. (See brackets.)

Question marks. Surprisingly many writers fail to notice that they've just written a rhetorical question, and mistakenly end it with a full stop. Or do they assume that because such a question (like this one) doesn't actually expect an answer, it's not a real question?

Quotation marks. When typing use double quotes as mentioned last issue, unless your publisher begs you to follow a different house style. Quotations within quotations get single quotes; quotations within quotations within quotations are probably a mistake, but it's back to double quotes again. (And so on.) Punctuation goes outside the quotes for isolated phrases or words, like 'this', but inside for speeches: 'Do it this way,' said Langford. (American usage differs.) In Grocer's English, quotation marks are used merely for emphasis. Discerning readers can thus enjoy placards saying things like *'Fresh' Lettuce,* which actually conveys that the word 'fresh' should be pronounced in tones of extreme sarcasm.

Semicolons. I am addicted to semicolons; readers may have noticed this terrible habit. Use them to link vaguely related sentences when complete separation with a full stop seems a bit too sundering. The decision tends to be a matter of personal style rather than grammatical compulsion. Downmarket newspapers will probably convert all your semicolons to full stops anyway, and then start a new paragraph after each full stop. This is supposed to make for easier reading – just as a meal is so much easier to eat when each potato is served as a separate course.

Spaces. The space is the most important mark of all, and the most abused. Of late I've seen spaces put immediately *before* full stops, commas, question and exclamation marks, semicolons, colons and right-hand parentheses – as well as immediately *after* left-hand parentheses. All these disgusting practices must stop at once. Nor will you be forgiven should you sleazily omit the space *after* the full stop, comma, question mark, etc. Some typing purists demand two or even three spaces following each full stop, but this remains wholly optional.

... Speaking of space, I've used up all mine. For further reading, consult G.V. Carey's *Mind the Stop* or Kenneth Tynan's substantially funnier essay on punctuation in *Tynan Right and Left.*

8000 Plus 36, September 1989

Occupational Diseases

Does your PCW give you nasty pains in the back? Suspicion eventually fell on my various computers when early this year I found myself groaning, limping, hurling myself out of bed screaming with cramp, etc.

Friends rallied round at once. Being my friends, they started by diagnosing kidneys wrecked by alcohol, went on to suggest that I was paying the inevitable penalty of being too tall ('Your disintegrating backbone just gets worse all the way to the grave now.'), speculated on loathsome viral ailments unknown to science, and hit bottom with merry hints (many of them from the other director of Ansible Information Ltd) about spinal cancer.

My doctor took a less alarmed view when he discovered how much time I spent hunched over word processors. It was the old problem of correct typing posture, which you tend to forget when running your fingers over something as effortless as a PCW keyboard. A contributing factor is that after all one's investment in computer hardware, there's rarely much spare change for mundane matters like office furniture. Rather suddenly it dawned on me that despite the above-mentioned Langfordian tallness, all my work for one magazine was being done at a battered little desk which my brother-in-law had used at the age of twelve. This is known as stupidity.

I'd better break it to you that grown-up desks with plenty of legroom are not available on NHS prescription. However, the investment worked well enough to make me recommend taking a critical look at whatever rickety washstand or tottering card-table currently holds up your PCW. With bad luck like mine, the result can be the kind of disk inflammation which Dave's Disk Doctor Service Ltd is not equipped to handle.

Here are further totally ill-informed health notes.

PCW Pink-Eye is merely a harmless optical after-image effect, whereby after long staring at a green screen, you temporarily see pale objects as pinkish. Immediate first-aid action consists of telling yourself loudly that this is *not* some frightful irreversible damage caused by dread VDU radiations. I have tried, and don't recommend, swapping each half hour between the 8256/8512 and a machine with an amber monitor. An alarming intensity of after-pinkness might indicate poor workroom lighting: even if you touch-type perfectly and think you look only at the self-illuminated monitor, excessively dim surroundings tire the eyes.

(Personally I advise keeping the monitor brightness turned well down. One of those mesh filters might also help, by eliminating reflection from the screen; but if your screen is a glittering riot of reflections and highlights there's probably something wrong with the arrangement of workroom lighting and furniture.)

PPC Finger, a more complex syndrome, results from the interaction of one's old-fashioned, metal Anglepoise lamp with the liquid crystal display of an Amstrad PPC luggable computer. After the 827 adjustments of screen and lighting angle required before you can view this wretched display even semi-comfortably, you'll have acquired several painful blisters from the hot metal shade. Cure: a midget fluorescent desk-lamp. More expensive cure: sell the PPC to an enemy and find a portable with a backlit display.

IBM Hernia needs no more equipment than one of those old IBM XT clones built like the legendary brick outhouse. Simply rearrange your office furniture on medical advice, pick up this machine without first dismantling it into the smallest possible bits, scream while putting it too hastily down, and seek more medical advice.

Paper-Align Jitter, a nervous affliction of the wrist muscles, begins to set in after the first fourteen attempts to get the PCW printer to roll in a sheet of A4 without tilting it just slightly out of line. Sprocket-fed continuous paper is the only known cure.

Write-Protect Fingernail occurs when, as always, no stout ballpen is to hand when you need to protect or unprotect the sort of disk requiring manipulation of a tiny, recessed and exceedingly stiff lever. (Warning to DIY enthusiasts: oiling this lever is not a good idea.) The resulting split and splintered nails can produce hideous side-effects if you ever idly pick your nose while brooding at the keyboard, but for the sake of the squeamish I will not go into detail about Write-Protect Nostril. Cure: try and stick to disks with sliding protection tabs, not the lever-action variety. Sorry about that, Maxell.

Daisywheel Despondency principally attacks 9512 buyers who thought they would be getting something much better in every way than the presumably cheaper and nastier 8256/8512. Agonizing bouts of existential dread and despair follow the discovery that of all those hundreds of fancy LocoScript characters, only a very few of the snazzier ones can be handled by a 96-petal daisywheel. Cure: pay extra for a matrix printer, or gloomily learn to fake exotic symbols – for example, type © copyright signs with brackets as (c).

The Bottom Line. Which aspect of literary health is the most fundamentally unsound? In the end, the long-serving professional writers who sit all day round at their word processors are most often heard to complain in embarrassing detail that it gives them piles. If you are seriously afflicted by these painful heaps of abandoned drafts and early print-outs, you should at once consult a qualified dustman.

8000 Plus 37, October 1989

Behind Closed Doors

September's magazine feature on the secrets of Locomotive Software prompted our ace reporter to detach himself momentarily from the bar and investigate the software company which is possibly the most obscure in the world: Ansible Information Ltd.

Ansible began life in 1984, and again in 1985, 1986 and 1988. For a company which has had no effect whatever on the working habits of nearly four billion people, its offices are surprisingly grandiose, consisting of two crumbling Victorian slum houses 45 miles apart. Asked how they can afford such palatial premises, head programmer David Langford quipped, 'We can't, but we have to sleep somewhere.'

Besides its basic commitment to unpronounceability, Ansible, as originally envisaged by chairman Christopher Priest, was to provide software solutions for unknown, obsolete computer systems which nobody owns or buys any more. 'While stealing computer time in the Oxford nuclear physics department, I learned to program an IBM 1130 by punching the cards with my teeth,' explained technical director David. 'Unfortunately this skill was less viable than hoped in the home computer market.'

Ansible's first commercial project was a system of pop-up menus which might have been a great success if restaurant owners hadn't objected to having slots sawn in their tables for the installation of this simple, spring-loaded device.

What complementary skills did production chief Christopher bring to the company? 'By then I'd written several highly praised though unremunerative SF novels in which shifting realities and hallucinatory narrative established a dreamlike state where no fact or interpretation seemed reliable. This left me ideally qualified to write industry-standard instruction manuals.'

How did Ansible enter the Amstrad market? Secretarial scapegoat David explained: 'As SF writers, we used to be forced to look at friends' terrible, badly-typed, unpublishable novel drafts. Then we noticed a change: more and more we were seeing terrible, unpublishable novels smartly produced on PCWs! This was an obvious pointer. Also, we had this idealistic notion about making obscene sums of money.'

An early Ansible product was the TYPO program, which could be run against LocoScript documents to introduce random spelling errors, misaligned letters, etc., thus catching the eyes of editors who'd grown bored with excessively perfect word-processed scripts. But these were early days for Ansible, and TYPO was withdrawn owing to a slight bug which in its first releases (up to version 4.79) could cause PCW monitors to explode.

Is the computing world anything like the directors' former haunt of SF writing? 'Oh yes,' replied switchboard operator Christopher. 'The combination of good reviews with low profits and huge tax demands is very nostalgic. We keep sales down partly by writing software for obscure jobs no one wants done – like indexing – and partly by our policy of not answering the phone.'

Our reporter was shown around Ansible's trophy room, and peered with revulsion into the glass case containing more than 47,000 pin-mounted bugs from early programs. On the wall are framed letters from computing giants Locomotive, WordPerfect Corporation and many more, all telling Ansible to watch it if they don't want to get sued.

Is it possible to explain Ansible Information's fabulous lack of success? 'I put it down to beards,' commented tea-boy David. 'In big-name software houses, male staff have peculiarly irritating beards – look at that horribly hirsute lot at Locomotive. Unfortunately my wife won't permit such a drastic revision of our public image.'

We followed the Ansible team through a complete day's work, beginning with intensive hours of oversleeping. Software boss David expertly showed how five minutes of making random changes in a program can quite often move the bugs around a bit, while public relations maestro Christopher shouted down the phone at multi-million-pound companies who as usual wanted a £29.95 software package but claimed total inability to write a cheque for such a huge amount in less than six months.

After a long discussion about the parentage of HM Inspector of Taxes in the company's nearby boardroom, known as The Plasterers' Arms, the mailroom supervisor (Christopher) and philatelic salivation operative (David) gave an exciting demonstration of how on a busy day Ansible often mails out enough software parcels to be counted on the thumbs of both hands.

Of which of its achievements, then, is Ansible most proud? 'Our manuals,' insisted technical authorship co-ordinator Christopher. 'We print fewer split infinitives and maintain a higher level of semicolons than almost any other doomed company of comparable size based in Reading.'

'I'd say our support service,' contradicted customer liaison assistant David. 'Within weeks of receiving a routine letter of complaint or death threat, we rush back a full explanation that the bug in question only appears to be so because they've misread the manual, and in fact doesn't exist, being instead a valuable feature requested by thousands of past users, which in any case results from flaws in CP/M or Amstrad's hardware.'

Why the name Ansible? 'We wanted the software to go ever so fast,' said nomenclature supervisor Christopher, 'so we stole the name of the fastest thing in SF, the instantaneous communicator in the novels of Ursula Le Guin.' Only later did they discover that it's an anagram of 'lesbian', which amuses their customers greatly and frequently.

Asked whether PCWs were used to prepare the manuals for their PCW software, both members of Ansible's product documentation section shuddered and said, 'Do you think we're mad? What do you use to produce *8000 Plus*, eh?' Our reporter made an excuse, collected his mac, and left.

Discussing proposals to write up this profile of Ansible Information as a four-page publicity feature packed with colour photographs, the chirpy editor of *8000 Plus* laughingly observed, 'Not on your nelly, Langford.'

8000 Plus 38, November 1989

• *There had recently been a closely similar feature on beard-ridden Locomotive, the vast software company which in those far-off days bestrode the Amstrad PCW world like a colossus, trampling jewelled thrones beneath its sandalled feet.*

Contracting Universe

Some people think becoming a professional author must involve a terrible initiation rite leaving lifelong scars: the brand of a red-hot keyboard, perhaps, or the outlines of three-inch disks tattooed on each buttock. Actually the true and appalling rite of passage comes when, delirious at hearing that Pemmican Publishing Ltd likes your masterwork, you're abruptly brought to earth by a horrible document called the contract.

This intimidating 'Memorandum of Agreement' tends to be printed on legal paper, hardly more flexible than tablets of stone and (maddeningly) too long for an A4 photocopier. With experience, one can suspect a psychological gambit: most literary agents keep standard contracts as word-processor documents because they're infinitely negotiable, while publishers prefer the printed look, to give the impression that they aren't.

The impulse is to sign at once for fear of losing your first big sale. Such action is invariably unwise. There's always the chance of a better deal.

You can pass the buck to your literary agent; if you have none, a book offer from a reputable publisher is excellent leverage when persuading one to take you on. Or ... grit your teeth and do your own negotiating. Don't trust my omniscience! A good starting point is the Society of Authors booklet *Publishing Contracts*, £1.50 from them (84 Drayton Gardens, SW10 9SB) or free to members.

Whichever your decision, you should read the contract yourself: though intimidatingly formal at first sight, each clause ought on examination to make some sort of sense. If not, ask your agent or publisher to translate.

The money arrangements tend to be straightforward (though beware of 'vanity' publishers who expect *you* to pay *them*). In exchange for various publishing and licensing rights, Pemmican Books undertake to pay you so much (perhaps negotiable) as an advance against royalties on sales, calculated as a percentage (not so negotiable) of the book's cover price. Certain subsidiary rights will also be covered, such as book club sales, translation, newspaper serialization – all potentially yielding loot for division between you and Pemmican (proportions highly negotiable).

Pemmican will undertake to do its sums regularly after publication, usually twice yearly, but will claim inability to write a cheque for royalties due until months after the accounting date. (Sometimes negotiable, to little effect.)

What about traps? Pemmican Publishing are not crooks (opinion negotiable after polling their authors) but want the best possible deal – and protection – for themselves. Here are some points to watch.

- Will the *copyright* be in your name or your pseudonym's? Any other arrangement is a Bad Sign, except in special cases of 'work for hire': for example, software instruction manuals are normally copyrighted by the software company, hundred-contributor encyclopaedias by the editor or publisher.

- How long before publication? The Pemmican agreement probably says 'within a reasonable time' from the contract or delivery date. 'Within eighteen months', say, is preferable. There's nothing more disheartening than indeterminate delays, especially if the book is at all topical.

- Will you see the final, copy-edited version of your typescript? Often it vanishes into the system and you know nothing until printed proofs arrive – a bad time to find that excessive tinkering has mucked up your book. The contract should specify that you get two sets of proofs for correction, and can keep one for reference.
- Does your book need illustrations or an index? Are photo copyrights owned by third parties? Pemmican will want you to pay reproduction/illustration fees; you want *them* to pay; you might end up splitting it 50%. Likewise for the services of an indexer if you can't face the job yourself.
- Is there a *reversion clause* (very important)? This ensures that rights return to you should Pemmican go bust, allow your book to drop out of print for more than (say) a year without scheduling a requested reissue, or remainder it.
- Is there an *option clause* giving Pemmican first refusal of your next book? This can be a pain: when Gastric Editions ask you urgently to do a lucrative book for them, you may not want a long, delaying ritual of first offering it to Pemmican. If Pemmican insist, there should be a time limit for their decision: six weeks, say. Having rarely written two similar books running, I've sometimes defused such clauses by allowing an option on my next 'of a similar nature'.
- What sweeping guarantees do you give by signing this contract? It's usual to declare that the book doesn't infringe copyright (e.g. by being plagiarized outright from John Fowles) and isn't libellous. Picky publishers may request disturbing extras, whereby you warrant that the book doesn't contravene the Official Secrets Act, and even, just recently, that it isn't blasphemous. The last in particular should be resisted at all costs. How can you predict what some maniac in Whitehall or Iran will retrospectively decide shouldn't have been published?
- What happens when they remainder your book? (When. Not if.) In the contract, Pemmican should promise to give advance notice that they intend to administer this humiliating kick in the groin, and to allow you first option of buying stocks for resale or Xmas presents ... at the low price offered to the trade. (My first remaindered hardback suddenly dropped to 50p when I queried the £3 asked.) Pemmican will happily agree to such provisions, and as a rule will then ignore them, knowing there's little you can do.

('The marketing division remaindered it without telling us,' you hear from editors. 'Copies were not moving fast enough out of the warehouse,' the marketing people explain, as though they'd been waiting for the wretched things to evolve legs.)

One last tip: distrust the common clause which goes, 'The AUTHOR shall forfeit his/her immortal soul for a period of not less than ONE ETERNITY from the date of this Agreement' ...

8000 Plus 39, December 1989

Electromagnetic Etiquette

Dear Miss Magnetic Media,

I've become acquainted with a famous author who nervously mails 'security copies' of his novel-in-progress to friends. This is great because you get the perk of reading the books before they hit the best-seller list. When his latest was published,

I naturally returned the disks with the draft versions. Back came a thank-you letter which I thought slightly chilly. Have I blundered?
– 'Worried.'

Miss Media replies: You have encountered a problem of etiquette so novel as to be absent from the guidebooks. See below ...

Dear Miss Media,
I'm a starving writer and recently made my first sale. As recommended by that bigmouth Langford, the print-out's covering page said: 'Text available on 3" disk in LocoScript or ASCII format.' The magazine was glad to avoid retyping, so all was well, except that months later they hadn't returned the disk. When I enquired, they sent it back sort of reluctantly and said something about being afraid of insulting me! Should I now feel insulted, and if so, why?
– 'Impoverished'

Miss Media replies: This embarrassment arises from friction between different social and financial strata in computing. It began when well-heeled organizations and computer owners chose to become uppity about returned disks. 'We don't ask people to send back letters so we can clean the paper for re-use,' one imagines them sniffing. 'We can afford endless new disks. Returning them implies that we're penny-pinchers. What an insult!'

Alas, a PCW user who mails many disks cannot afford this haughty contempt for bank managers. The big-time attitude was formed by the traditional IBM 5¼" disk (which in bulk costs as little as 20p) or the cheapest 3½" equivalent (perhaps 50p), rather than the 3" Sugar Special at a minimal couple of pounds.

Miss Media suggests that much heart-searching might be avoided if senders of material on disk were to inscribe the label *No need to return this*, or alternatively, *Please return after copying files – thanks*.

Those objecting to the latter plea should contemplate the virtues of Green policy. Causing correspondents to buy new disks uses up resources; recycling old ones doesn't. Conversely, if you're so eager not to insult others, do you throw away their disks to avoid demeaning yourself (scandalous waste!), or furtively re-use them? Be honest, now.

Dear Miss Media,
Are you aware that cardboard mailing boxes for floppies cost more than new disks? Ecologically it's better to throw disks away than mail them back.
– 'Megacorporation'

Miss Media replies: Miss Media was addressing the long-suffering PCW owner, whose sturdy little disks can be safely posted in used jiffy bags, or even ordinary envelopes (cardboard stiffeners are advised). Doubtless there's small hope of persuading corporations, no matter how Green, to soil their squeaky-clean image by mailing all disks with cardboard protection cut thriftily from supermarket boxes. A pity.

Twice each blue moon, a disk is scrambled or cracked in transit. Accept this as our kindly post office's way of making you grateful for *(a)* keeping battered old disks for mailing; *(b)* retaining backups.

Dear Miss Media,
This public domain disk has me foxed – there are no instructions, and I can't load the file READ.ME which might explain things.
– 'Baffled'

Miss Media replies: To book-lovers, that filename evokes some unpublished Lewis Carroll fragment in which Alice discovers a small text file called READ.ME and finds on doing so that it magically turns her brain to mush.

The READ.ME (or README.DOC, or READTHIS.NOW, or whatever) information might be in LocoScript format or 'plain ASCII text'. Those accustomed to CP/M or other machines' MS-DOS will expect ASCII, and react by entering TYPE READ.ME at the CP/M prompt. Hardened LocoScript users naturally attempt to peep with E for Edit.

If it's ASCII, LocoScripters should create a new document and load in READ.ME via 'Insert text' (**f7** 'Modes' in Loco 1; **f1** 'Actions' in Loco 2).

If LocoScript ... might it be suggested to perpetrators of READ.ME files that they use Loco 1, which everyone can read, and not Loco 2, which frustrates unregenerate Loco 1 users? To comfort those who try CP/M and TYPE, a message in LocoScript's 'identify text' (editable through 'Modes'/'Actions') is a shrewd ploy. This text, preceded by the letters JOY, is what TYPE will display when READ.ME is a Loco file; its 30 characters allow room to declare, 'Read me with LocoScript, clot!'

Miss Media retains an old-fashioned preference for clearly printed instructions on archaic paper.

Dear Miss Media,

I sent my novel on disk to this writer I know, but she says she's scared of inserting strange disks for fear of viruses. How can I reassure her?

– 'Plague Vector'

Miss Media replies: Miss Media can only applaud the brilliance of your friend's excuse for not reading your ghastly book, while answering with enormous regret that it's unfounded. Were a PCW virus ever to emerge, it would be spread by infected 'start of day' disks, or perhaps '.COM' program files in CP/M. Using your own LocoScript to examine alien document disks is always safe.

Dear Miss Media,

Har har CAUGHT YOU OUT! You said reading disks was ALWAYS SAFE what about this then. I pull back the shutter, put on a streak of superglue and sprinkle with fine carborundum. Anyone reads it, they need a NEW DRIVE, brill eh?

– 'Smartass'

Miss Media replies: There is a time for measured considerations of etiquette and there is a time for petrol-soaked blazing crosses to be hurled through windows. In your case one is reluctantly compelled to the latter course.

Careless tea- or coffee-drinkers are warned that a dried spill on the disk surface may have almost as exciting an effect as carborundum.

8000 Plus 40, January 1990

Post-1989 Miscellany

Having recently returned from a longish trip to the northwest USA as guest of a science fiction convention, I planned to write an in-depth analysis of the PCW market over there. Here's a complete transcript of my research notes.

Diligent Investigator: 'How popular is the Amstrad PCW in these parts?'
Many American Literary Persons: 'What the hell's a PCW?'

Half of them own IBMs and half use the incredibly expensive (at least in Britain) Macintosh. Although the statistics show that a tiny band of American PCW users does exist, they seem as difficult to trace as that statistically common family with 2.4 children.

Over there, it appears, the PCW came on the scene too late: makers of cheap IBM clones had carved up the penurious end of the market in a bloodbath of price-cutting which we didn't see here until much more recently. End of in-depth analysis.

8 or 9?

The cover says February, you're probably reading this in January, and I'm writing at the height of the Xmas horror: we SF fans who've spent our lives reading about time distortions always feel right at home in publishing.

So, as I write, fond relatives are still ordering software for the family PCW addict and failing to specify whether the computer is an 8256, 8512 or 9512. *[Or, later, a PcW 9256 or 9512 Plus.]* Let's draw a veil over those who fail to mention that it's actually, ahem, a CPC 464. Indeed this happens all year round, and often it's the computer owners themselves who don't choose to reveal intimate facts like model numbers to strange software dealers.

Of course it's frustrating to receive a disk which won't start up your machine or can't even be read on it. No disk which starts up the 8256/8512 will do so for the 9512 or vice-versa, and inevitably the 9512's 720K disks remain inscrutable to 8256 owners. My own outfit, Ansible, reckons that the best guess is to send a 180K disk for the 8256/8512, with a note explaining that 9512 owners should either *(a)* copy the programs to a 9512 CP/M start-up disk as so simply and beautifully described in the manual, or *(b)* send it back for recopying. To save their supplier from idle hours of staring out of the window, most people choose *(b)*.

So ... do remember to mention your machine model when ordering programs or dropping tactful hints to relatives. I admit that sensitive, intelligent *8000 Plus* readers rarely need this exhortation. Ansible's main quarrel is with uncultured chaps who forget to state the machine number, sign the cheque, etc., and then make witty remarks like: 'Don't waste my valuable time with questions – it's *your* job to get these things right.'

Disk Full, or Not?

Another problem I have to explain from time to time concerns the exact meaning of 'free space' on a disk. The awkwardness doesn't arise if you only ever use LocoScript, or if you never use it at all, or if you keep separate working disks for LocoScript documents and other files. But if you put Loco files on the same disk as those produced by CP/M programs (perfectly legitimate, but see below), you'll sooner or later meet the Great Disk-Full Paradox.

'100k free', says the Disk Manager menu in LocoScript. Who could disbelieve it? 'No room on disk,' insists an otherwise apparently reliable CP/M program. 'ERROR: DISK WRITE NO DATA BLOCK - A:DOCUMENT.$$$', reports CP/M's dear old user-friendly PIP, which is its way of saying, 'No room on the disk, squire.'

Old hands at the PCW game will already be wearing knowing smiles. This is all to do with the dark mysteries of LocoScript 'groups'.

Groups started as a CP/M idea for letting a number of different users, or groups of users, each have their own 'private area' of a disk. When you type DIR in CP/M you normally get a directory of Group 0 (zero), and most people vaguely think that only LocoScript can see Groups 1 to 7.

In the old days, user number 1 would type USER 1 in CP/M, and the A> prompt would change to 1A>, and he or she would have access to all the Group 1 files. In fact this still works in CP/M on the PCW. Enter USER 7 and then DIR, and you get a listing of the Group 7 files.

CP/M doesn't stop there; the USER group numbers go up to 15. This is where complications appear. Loco uses Groups 8 to 15 solely for Limbo files. When a Group 0 file is erased or edited, the old version is simply relabelled as being in Group 8 and can (as we know) be hauled back. Likewise Group 9 is the limbo for Group 1, Group 10 for Group 2 and so on.

The Limbo groups aren't counted in LocoScript's calculations of free disk space: files there are living on borrowed time, and Loco will erase them 'properly' the moment it wants the space.

To CP/M, though, Groups 8 to 15 contain ordinary files with an equal right to live. A much-used Loco disk will have a crowded Limbo area and will look pretty full to CP/M.

These different views of the higher-numbered groups explain the cryptic warnings in manuals: stick in a CP/M disk containing a needed file which happens to be in Group 15, and LocoScript might liquidate it as an expendable resident of Limbo. But who these days uses the groups, except in LocoScript? A few lucky hard disk owners, perhaps.

Finally, the obvious solution. When you want a CP/M program to write text to a Loco document disk (perhaps for later 'Insert ASCII' inclusion into a Loco file), first select Loco's 'Show Limbo' option and erase some suitably unwanted-looking Limbo files from the revealed list.

Another solution: you could always have saved the CP/M program output on a freshly formatted disk, but where's the technocratic fun in that?

8000 Plus 41, February 1990

Swings & Roundabouts

An old friend called to bewail his feelings of computerized obsolescence. His dearly loved Amstrad PCW is worn and tatty, and people (he says) laugh at him for clinging to a machine bought three whole years ago. 'Is it time for a change?' he asked with a technophilic gleam in his eye. 'State of the art. High tech. Fast lane. Maybe in matt black to match the music centre ...'

Well, possibly not black – the business computer industry seems committed to pale grey or beige cases, ingeniously toned to highlight the dingy evidence of actual use. (Why is the filthiest key on my PCW the J?) But a faster system does seem to be a lure.

I cheered him by making up and quoting the hardware version of Parkinson's Law, which is that your expectations of the most fabulous new machine will expand until, sooner than you think, you're grumbling at its limits.

One instance of the law is known to writers of this column as Terry Pratchett's Insight: 'There's no such thing as a fast computer after the first three days.' By this time, the super new machine has become normal (i.e. faintly if not quite annoyingly sluggish), while your old one is now so absurdly slow as to be unusable.

Indeed, strangely enough, I find my IBM word-processing system doesn't seem that much astonishingly faster than LocoScript. The IBM may run at ten times the PCW's speed, but against that is the fact that the software I use can handle a full-length book as a single file. In accordance with the Law, the eager writer does exactly this, ungratefully takes the facility for granted, and spends his or her time moaning that as a consequence, moving from end to end of a document occupies the same sort of interglacial aeon familiar in PCW legend....

Optional Extras

'You mean it *doesn't come with software?*' said my pal in bogglement, and fainted when I explained that the market leader amongst IBM word processors cost £425 plus VAT, admittedly negotiable to much less by avoiding posh computer dealers with nice suits and carpets.

Indeed, the pricing policies of multi-national software companies are a real running sore in computing. I still gnash my teeth at being an authorized dealer for a package we'll call ExpensiveWord, and finding that the suppliers *(a)* flogged software in bulk to mail order box-shifters who could make a profit selling them at less than the discount price quoted to authorized dealers; *(b)* in answer to enquiries, advised prospective customers that 'the going rate' was less than we as dealers could buy it for, let alone sell it for.

But I digress, as usual. Although it's barely possible to get by with public domain stuff, anyone seriously planning to invest in an IBM system should be careful to budget a few hundred for word-processing software alone.

Other extras include a printer, rarely part of the deal, and a printer cable, and perhaps a complicated mass of communications software, interfaces and cables to allow transfer of documents from the old PCW.

I fell into the 'extras' trap this year when tempted by a special offer of a DIP Pocket PC, better known under its 'Atari Portfolio' alias. It came with batteries and built-in ROM software, and is the first portable computer which both has a real operating system and *does* go in the pocket instead of being luggage.

Immediately, it didn't work. Extra item number one was a fresh set of batteries. Then I ran into software bugs (never try to save an empty document, for example) and had to reset the machine. This meant losing everything in memory. Extra number two was therefore the equivalent of a disk: a battery-backed memory card whose price made my eyes water.

Extras three and four were needed to transfer stuff to my desktop computer: an interface box and a peculiar cable which a Tottenham Court Road shop first couldn't believe in and then had to make up specially.

Reading the small print that came with the interface, I was instructed to buy extra number five in the form of a mains adaptor, since transferring files is such a frightful drain on the batteries....

It's a sweet little machine, though, and when I've dredged up a technical manual and/or another sort of cable (extras number six and seven), I'll report on how handily it can work as a roving accessory to the PCW.

One safe bet is that although the Pocket PC behaves like a tiny IBM, the still-delayed IBM LocoScript will *not* work usefully with its strictly limited 40x8 LCD screen.

Yuppie Bait

Another friend has just bought an Amstrad PPC: imagine her delight at finding it came with a free cellphone. Now she too could chat with her stockbroker while whizzing along the M4 fast lane and working the fax machine with her other hand.

Once again, there were some extras:

'Connecting' the free cellphone costs £60. Line rental is £25 per month, plus call charges. Insurance against loss is compulsory, at £3.50 a month. You pay a further £3 a month 'invoicing charge'! And none of the above includes VAT. All payments must be via a direct-debit authorization allowing the cellphone mob to take what they like, when they like, from your bank account. 'Credit worthiness' is a condition of contract, and as a display of trust in this quality they demand a deposit of £200 against all the above.

As my software partner said to me, 'We're in the wrong business, mate.' It certainly sharpened our irritation at motorway punters yakking obliviously away on their wretched phones while overtaking: now we *know* they're all vastly richer than us....

8000 Plus 42, March 1990

Stalking the Wild Editor

8000 Plus is read by hordes of aspiring authors, all eager for insights into the strange ways of the editorial mind. In hope of gratifying this prurient interest, I polled SF/fantasy editors and asked what they like least about word-processed submissions. Out of curiosity I also asked what percentage of such material was recognizably produced by LocoScript. Now, in a flourish of inadequate sampling, I reveal all.

... Or nearly all. I was soon reminded that editors are constantly overwhelmed by their work of sifting the million or so books submitted each year, to decide the 50,000-odd that are published. Gollancz was busy changing editors during my survey, and Pan was busy relaunching its fantasy, SF and horror lines, while my contacts at NEL and Futura must have been busy in the pub.

Malcolm Edwards of Grafton Books began with a careful distinction: 'I'd think that of the unsolicited UK submissions, a quarter to a third are done on the old Amstrad. These days, published authors tend to have a better class of printer, at least, and often have cheap IBM clones.'

Note that all unpublished authors, barring a few famous politicians, media stars and murderers, will land on the 'unsolicited' heap – alias the slushpile.

Malcolm Edwards's unfavourite things are: '(1) authors who think that the ribbons last as long as the computer, and that the extra wait involved in printing out NLQ pages is more important than my eyesight; (2) the fact that authors who no longer have to retype everything seem to use the time saved in writing longer books instead of using it to write better books of the same length.'

Deborah Beale of Century Hutchinson also has me nodding in agreement: 'It drives me *mad* if I'm sent continuous paper print-outs, where the author hasn't bothered to separate the pages. Also I loathe right-hand justification. The monotony of such a neat page sends my eyes funny, and often, sort of like the white-noise hiss of air conditioning, sends me to sleep.'

Continuous paper is simply difficult to manage, especially for editors who nobly catch up on their reading in the train, armchair or pub.

David Pringle of *Interzone* bewails the side-effects of this magazine's increasing circulation: 'The quantity of unsolicited submissions still seems to be going up and up. While one ought to be grateful, it's true that they can be such a pain in the [*now, now – Ed.*]. Perhaps my main complaint is the obvious one that so many people don't bother to obtain a copy of the magazine before submitting. There seem to be endless thousands of aspiring SF writers who get all their information from the *Writer's and Artist's Yearbook*, and nowhere else....'

The point is that every magazine has its own indefinable 'feel'. If you write a thrilling story headlined RAMBO VICAR IN SEX ROMP MERCY DASH, its undeniable relevance to religious matters may not impress the editor of *Church News*. Likewise, a scholarly and literate article on UFOs would obviously be rejected out of hand by the *Sunday Sport*. Unless you like wasting postage, it's vital to study your market.

David Sutton of *Fantasy Tales* castigates the evils of 'cramming a 30-page story folded into four in a standard letter envelope ... text running across whole width of page with no discernible margins ... no indentation of paragraphs ... no return postage!'

This reminds me of once receiving a carefully unsealed envelope which had been laboriously embossed with several artistic impressions of 10p pieces. The letter inside achieved a final stroke of unconvincingness with: 'PS, I hope your postman does not steal the return postage enclosed....' Full marks for enterprise and economy. The MS went into the bin.

Another niggle from David Sutton: 'I occasionally get a MS with the title page set out as though produced on a desktop publishing system, with a flourish of different typefaces for title, author, word count, address, and computer system used! There's no reason why I should hate this except jealousy....'

That is, editors want functional, readable text, and react badly to ornaments which are the typographical equivalent of submitting work on deckle-edged lavender paper impregnated with perfume.

Wayne of *GM* magazine doesn't confine his gripes to amateurs: 'Writers either can't count or badly miss the option of a word counter – especially on Amstrads. We informed readers that we were after short stories of up to 1000 words absolute maximum, to take up one magazine page. What happened? The average length of Amstrad-composed material submitted to us was around 1800 words!

'When we asked "name" authors to submit short stories of no more than 4800 words, they too went over the top. Most averaged around 6500 and one established author exceeded 8000. The excuses: the amateurs are honest and state that excitement plus lack of a word counter made them overestimate. The professionals put it down to artistic licence.'

If you don't fancy investing in word-count software, it's worth hand-counting one big chunk of typical print-out in your normal format, to give a basis for future estimates. Be careful not to include scanty pages of dialogue – 'What?' he said. 'You know,' she snapped. 'No I don't.' 'Yes you do.' 'No I don't ...' – unless of course you write like this all the time.

Finally, most of the editors in my amazingly wide-ranging survey felt that around 25% of their submissions were detectably matrix-printed on a PCW. *GM*, being largely a specialist market (SF/fantasy games) with strong amateur involvement, was exceptional at 65%.

To draw statistical conclusions from this sample would be mean. Instead, please consult Havelock Ellis's *Psychology of Sex* to learn the standard deviation.

8000 Plus 43, April 1990

Software Deathwish

If it weren't a violation of copyright, I'd be tempted to suggest that every Amstrad monitor should come stencilled with the large friendly message DON'T PANIC. So many people seem to prostrate themselves before their PCWs in an ecstasy of panic, that a time traveller from ancient Greece could reasonably deduce the machine to be a small household shrine of the great god Pan.

This state of terror resembles the Examination Dream suffered in later life by so many victims of education. You're confronted with a nightmare list of questions which cannot possibly be answered, even if you still knew anything about the forgotten subject. Cold fear takes you by the short and curlies, your ribcage suddenly becomes three sizes too small, sphincters begin inexorably to loosen ...

For all too many PCW owners, this whole complex of terror and mental paralysis is instantly conjured up by the CP/M prompt message (readers of a nervous disposition should skip what follows): **A>**.

Software people who try to explain things over the phone are all too familiar with the symptoms of panic. One is a sort of wild, despairing haste, leading to exchanges like this.

Expert: 'Now I want you to do exactly what I tell you. First, put in the disk with MASTER written on it. Then −'

Voice on Telephone: 'Yes yes I know that bit,' *tap tappety tap*, 'I'll just start it up,' *tappety tappety*, 'and now Oh *God!* It's just not working, it keeps saying question marks, I'll have to turn the thing off and start again,' *CLICK*, 'Sorry, *which* disk did you say?' *tap tappety* ...

Expert: 'Stop! Stop hitting keys and listen!'

A related symptom is that sufferers tend to get stuck in a loop, repeating some quite simple error (wrong disk in place, or a command slightly mistyped) under the firm conviction that they are following the exact procedure in the instruction book.

Expert: 'Let me go through this keystroke by keystroke. Type these letters....'

Voice on Telephone: 'That's *exactly* what I've been doing all ...' (Long pause. Very querulous tone.) 'It never did *that* before.'

I sympathize with the victims, who have usually worked themselves into a ragged emotional state before calling the 'expert'. Unfortunately, frayed nerves can lead to the helper's getting a hard time.

Expert: 'Let's see if all the right files are there. Type DIR and tell me what −'

Possible Reply 1 (splenetic): 'Look, I know what disk I put in, and what's more I know all about listing directories in CP/M, so you needn't treat me like some stupid kid, thanks very much.'

Possible Reply 2 (half-sobbing): 'Oh dear, you go on about typing DIR and all this technical stuff, I just can't understand any of it, why won't you stop trying to blind me with science and just tell me what to *do* ...'

This last response approaches the final stage of panic which I think of as 'software deathwish'. Stand back! I am about to commit amateur psychology. What seems to happen is that the panic victim becomes emotionally committed to *not* understanding, to *not* solving the problem. The buried reasoning might go roughly as follows:

'I'm not an idiot. But I'm totally baffled and frustrated by this software. It *must* be incomprehensible. This bloke who keeps saying it's really quite simple is obviously lying through his teeth. I'm not going to listen to him. (Besides, if it truly *is* simple, what does that say about me? Better not think about that.)'

Sometimes, trying to help callers who have reached this state, one is left with the uneasy feeling that immediately after hanging up they plan to hurl the PCW from a high window, and to jump right after it.

This being a far from perfect world, I don't say that all software is dead easy to learn (much is idiosyncratic) or that all experts are helpful (the most likely response to a call for support is 'Just putting you through', followed by twenty groan-laden minutes of Muzak). However, here are some hints on how not to make things worse by succumbing to panic.

• Allow yourself time to master new programs. Non-swimmers do not as a rule leap into the Channel, point themselves towards Calais, and reckon on picking up the techniques of breast-stroke as they go. Just so, the way to learn that new spreadsheet is not to try and move your complete business records system to it on the day the tax return has to be posted.

• Check all software if you change computers; there's nothing more panic-inducing that the sudden discovery that a familiar program has stopped working. That roof-rack you bought for the old Reliant Kitten will have to be replaced now you've traded it in for a Rolls: the same goes for all 8256/8512 start-of-day disks if you move to a 9512. But the pink fluffy dice from the Kitten are Rolls-compatible and may be transferred – as can most 8256 programs that run with CP/M.

• When ready to scream with rage, remember the magic mantra 'RTFM' which constantly trembles on the experts' lips. This stands for 'read the effing manual'. Just this once, forget all the short cuts which you 'know' are OK, turn off your imagination, and follow the instructions step by laborious step. If the program *still* doesn't work, you now have my permission to scream.

• After screaming, stay clear of the wretched computer until your pulse rate is back to normal.

• When you're as experienced as I am, you'll meet problems with a tranquil ... oh *sod* it, another missing address mark, and that's my only copy of the column, and the deadline is ... *AAAARGH!*

8000 Plus 44, May 1990

UFO Follies

It's a great tradition of *8000 Plus* that when you delve into the most arcane fields of human endeavour – toad-sexing, or science fiction fanzines, or societies for the preservation of fruit-bats – you'll sooner or later find someone labouring away at the paperwork on a PCW.

Therefore it came as no surprise when Jenny Randles, who is apparently the only full-time professional UFO researcher in Britain, mentioned that her books like *Abduction* (Headline paperback, 1989) are indeed produced using LocoScript 2 on a PCW8256.

At first I was leery of these researches, since UFOlogy can seem such a dubious subject: SPACE ALIENS TURNED MY PCW INTO AN OLIVE, and all that. It was with great relief that I found Jenny to be a leading member of what purely for the sake of argument I'll call the non-loony school of UFO thought.

Here in Britain, UFOlogists of this school suspect that the weirder reports from honest-seeming witnesses (close encounters, abductions, etc.) result from strange mental states. *Not* madness or DTs, but something more like lucid dreaming or those 'out of the body experiences' which seem so real to sufferers. This is fascinating stuff which might well provide light from an unexpected angle on how the human brain is wired – and how, like a computer, it can go into unintended loops or even, temporarily, crash.

Unfortunately there are problems in holding views like these. The first I could guess; my jaw dropped when I heard about the second.

The first snag for a sober researcher is that, as Jenny confirms, this kind of UFOlogy won't make you rich. 'My 14 books normally clock up about 2000 sales in hardback, with rare paperback excursions – giving my bank balance a status that even Argentina won't envy.' What sells is hyperbolic stuff about tangible alien spaceships full of little kidnappers with enormous eyes and faces made of putty. This is why UFOlogists in America, where this 'extraterrestrial hypothesis' dominates, tend to be very much wealthier than our home-grown ones.

Personally I have strong opinions about America's best-known UFO pundit, the dreaded Whitley Strieber of *Communion* fame. Jenny will not discuss him at all, owing to the second snag in being an unsensationalist researcher who disputes theories of aliens. This is that moneyed Americans are quite ready to go to court over these issues.

For example, Jenny's and Peter Warrington's *Science and the UFOs* (not a bestseller) is acknowledged by Strieber as an influence: almost immediately after reading its account of 'UFO abductions', he is supposed to have started remembering the similar experiences whose highly coloured write-up in *Communion* made him rich. When in a radio interview Jenny made the obvious joke about this sequence of events, she was threatened with a libel suit.

Here I must declare an interest. Strieber's new novel *Majestic* devotes two pages to a detailed rehash of the fictional UFO story in my own 1979 spoof, *An Account of a Meeting with Denizens of Another World, 1871*. I was not asked permission, nor offered a fee for the use of my original creation. But I'd better not make jokes about it on the radio, had I?

Worse is to follow. An even more bizarre American outgrowth of UFOlogy is the cult of the 'MJ-12 papers' – dodgy-looking documents which are supposed to be leaked US government records. They tell a tale of crashed UFOs, autopsies on little green bodies, global cover-up, and much more. This is supposed to have been successfully kept secret by every US administration since the 1940s! I myself support the document experts who reckon the papers are blatant forgeries, and who quote strong evidence for this view ... but I'd better not say so too loudly.

You see, Jenny said so last year. It's a nasty story, offered here as a warning on the perils of talking to the media – as, in hope of pushing our books, all we writers sooner or later must.

In brief: the *Manchester Evening News* announced (uncritically) a public meeting in which Stanton Friedman, the leading US guru of MJ-12, would preach to the converted. Jenny, who lives nearby, felt she had a responsibility to give the opposing viewpoint. This was dismally written up by the paper as something which could be taken as a personal attack (with jazzy paragraphs starting ZAP! and POW!). Although Jenny complained at once of being misrepresented, the MJ-12 crew issued writs.

This is where I boiled over. For producing the offending piece, the *MEN* is being asked for £500, and its reporter for nothing. For expressing her dissenting view in what's *supposedly* a scientific debate, and despite having no control over the distortions which appeared, Jenny is being sued for £10,000.

Which could make her original letter to the *Manchester Evening News* the most expensive brief document ever to be printed out from Loco 2 on a PCW....

Surely this is outrageous. As Jenny says, 'This sort of tactic has no place in serious debate on controversial issues ... it must be stamped out. This is why I have an obligation to fight on despite the horrendous difficulties of doing so.'

Writers shocked by what they see as persecution have set up a defence fund to help fight this expensive action. (I'm one of them.) Echoing the legendary fund-raisers of *Private Eye* magazine, it's called MJ-BALLS. Cheques made out to this worthy cause are welcomed at 17 Polsloe Road, Exeter, Devon, EX1 2HL. This is a serious appeal.

As for Jenny Randles, her final, rueful comment to me was: 'Perhaps the PCW tempts one into trouble because it's so easy to respond quickly to points you dispute – a side-effect your readers mightn't have contemplated!'

8000 Plus 45, June 1990

• *As usual, the money ran out before the case got to High Court (most 'libel' cases could be rapidly settled by the common sense of magistrates, but our wonderful English rules don't allow that). The MJ-Balls fund eased the pain of the inevitable settlement out of court. My thanks again to all who contributed.*

Amstrad PPC Blues

Like hordes of Amstrad Portable owners, I use the thing solely on the mains supply: it eats batteries with hoggish enthusiasm. One must harden one's heart against the constant plaintive requests 'Please set time and date' and 'Please fit new batteries'. There's a certain pleasant olde-worldeness about the resulting file dates, which indicate that I did stupefying amounts of work in the early hours of Tuesday 1 January 1980.

Last month, this idyll ceased when the external AC adaptor went quietly dead. After the usual checks on plugs, fuses, and whether a multimeter could detect any flicker of life at the low-voltage connection, I groaningly turned to the manual. It was time to wrestle with another Alan Sugar index.

Over the years, book publishers have developed a sort of tradition that the index should appear right at the back. What you find at the back of the PPC manual is a great wodge of previously unnoticed software licence agreements, full of warnings against making more than one copy of any supplied software (what, you're allowed only one start-of-day disk?).

The index is concealed just before this. After looking under 'AC', 'adaptor', 'mains', 'power' and 'supply', you begin to try desperate long shots like 'low voltage', 'DC', 'plug', 'fuse', 'connector' or 'burnt out'. No luck. Mains power supply problems are eventually found on a page indexed under 'troubleshooting, DOS', a surprise to anyone who thought the PPC's DOS operating system was a piece of software. The technical advice given is, 'consult your dealer.'

After an abortive wrestle with the duff power supply's long and hard-to-remove screws (which, when loosened, left the box still sealed shut), I found a unindexed warning in capital letters on page 8, saying DO NOT REMOVE ANY SCREWS. Giving in, I let my dealer break it to me. '£40 plus VAT, squire. It'll take at least a week.'

Mysteriously, they found one in their storeroom two days later and charged only £25.94 altogether. My statistically unsound projection was that, on the evidence to date, the running expenses of a PPC will include some £8 a year for new power supplies.

Meanwhile, if you have the same PPC manual as me, the place to look for insufficient information about the power supply is under 'setting up' on page 7, cunningly indexed as 'setting up your PPC, 10'. One urgently emphasized bit of advice is to disconnect the mains plug when not using the PPC. I suddenly remembered I'd accidentally left it on all weekend. Does anyone do a bolt-on cooling fan...?

• *This is a rare unpublished piece – part of a column vetoed by the current editor under the newly revealed Rule 42, 'Thou Shalt Not Mention Other Amstrad Computers In This Magazine.' Other parts were expanded into full-sized columns below.*

Dangerous Corners

One nice thing about being a small computer company is that, depending whom you want to impress, you can legitimately call yourself Chairman, Managing Director, Head of Programming, Marketing Consultant, Chief Buying Executive, and indeed every prestigious title not currently wanted by the other director. The snag is that you – that is, I – have to *do the work* of all these imposing functionaries.

It was as Buying Executive that I met a fascinating example of real-life logical paradox. One gradually gets used to the pitfalls of trade price lists, such as having to detect by telepathy whether the figures include VAT (usually not, except sometimes) and whether '5% discount for cash with order!' applies to cheques (yes, but expect irritating delays while the cheque is tested for bounciness). The new snag came from bulk discounts.

We wanted CF2 disks. The minimum bulk order from a company which shall remain nameless worked out at over £100. Which was fine until I noticed that when you ordered more than £100-worth, the price per disk dropped. Great! I did the sums again, using the lower figure. Oops.

Bertrand Russell would have loved this. At the price for orders less than £100, my disks cost over £100, and at the price for orders over £100, they were only ninety-something, which raised the price to the higher rate for orders below £100, so ...

A systems analyst would probably suggest, after pocketing a huge fee, that the neatest solution here would be a flat charge of £100 for every disk order in that dodgy region where the price keeps oscillating. Being a cynic, I suspected that the vendors might prefer avarice to logical elegance, and played safe with a blank cheque marked 'not more than (the higher price)'.

This is how, in the world of computers, it can take half an hour of head-scratching to make one purchase. It also shows aspiring programmers how innocent-looking rules may lead to dangerous instabilities. The places to watch for problems are at a situation's edges or corners; for the disk purchase, the awkwardness comes where the price per disk veers in a rather ill-defined way to the lower rate.

Glitches often happen at extreme edges, at zero and infinity. That zero disks cost zero pounds isn't a problem. However, as a book reviewer I often receive parcels of SF costing nothing: when I flog the unwanted rubbish at 50p a copy, my percentage profit per volume is 100 times 50p divided by zero, which is guaranteed to boggle any business records program. Unless, of course, some programmer has incorporated the tactful message, 'Division by zero? You can't do that there here.'

It wouldn't happen in real calculations? Think of this: a company, probably mine, makes zero profit in 1990. Whatever profit or loss it makes in 1991 will, if you're fool enough to express the change as a percentage, be infinitely better or worse than 1990's.

Another example which rather embarrassingly comes from real life: wearing my Chief Programmer's hat, I wrote some software which gulped LocoScript files and did nameless things with them. It was only interested in the text, and ignored headers, print controls for underlining, and so on.

But one chap's document made it seize up completely, leaving me baffled until I realized I'd failed to consider the extreme case. There *was* no actual text, and my program was no good at processing LocoScript files containing zero words.... After each word, and only then, it checked whether it was at the end of the file. The cure was to have it check *before* each word.

(I still haven't allowed for awkward folk who produce Loco documents of infinite length. As soon as someone announces an add-on hard disk with infinite capacity, I'll have to rethink that program again.)

Another time when I had to worry about the treacherous corner at Point Zero was in doing the astrophysics for an SF story. The easiest way to work out certain orbits, and the time it took before spaceship A collided with massive object B, seemed to be to simulate everything on the computer and let Newton's laws of motion and gravity take their course. (Forgetting Einstein's for the moment.)

Unfortunately the program tended to blow up shortly before printing out its graphs of what had happened. The gravitational attraction between A and B doubles each time the distance between them halves. For the first rough simulation, I hadn't bothered to allow for these objects actually having any *size*: as their separation shrank to zero the acceleration soared towards infinity, the figures got vaster than the computer's registers could comprehend, and the program pointed a last accusing error message at me as it died.

Obviously, if one object is little and the other a planet, the program should stop calculating and simulate a loud bang when the separation between their centres drops to whatever the planet's radius happens to be. Problem solved – or only evaded? Suppose there's a tunnel right through the world and our spacecraft plunges in ... does it meet awkward infinities at the centre?

This is interesting. It turns out that, as with the disk price list or the difference between water at 99 degrees and 101 degrees, the rules change when you cross an invisible line. Once below the surface, instead of growing by the inverse-square law, the force of gravity now decreases sedately with distance from the planet's centre, and of course reaches a neat zero when you get there. Sanity returns.

Before turning this corner, the calculations had overflowed because of dodgy approximations and not thinking the problem through. *Mea culpa.* When I'd put it right and cranked out my distance/time graphs, I went around for days being smug and calling myself the company's Astrophysics Director. No one seemed to believe me.

Column 46, *8000 Plus 46*, July 1990

Aftermath of Glory

When aiming these columns at aspiring book writers labouring on their PCWs, I may sometimes give the impression that heartache ends once your contract is signed, your advance banked, and your book in print. There are still grim days ahead.

What awaits you is the grimy face of marketing, the dark side of the sales force. The publisher's publicity department will arrange for you to promote your masterwork on local radio ... local, that is, to somewhere incredibly remote. After years of making such trips, many authors suspect that it's wiser to stay at the keyboard or the day job rather than traverse most of Britain to be patronized for about five minutes by a low-browed DJ who has neglected to read the book. 'So, er um Mr Rushdie, you've written these er wossname *verses* about, about Satan, does this mean you're interested in heavy-metal rock?'

As for the other promotion you'll get.... Once I was a guest of honour at the World SF Convention on one of its rare visits to Britain, and my latest publishers were offered lashings of free publicity, with the prospect of selling stacks of copies to the thousands on thousands of SF fans being indoctrinated with my alleged wonderfulness at the event. So shrewdly did they seize this opportunity that not one copy of my novel was available throughout the entire convention.

Indeed, all your relatives, friends, acquaintances and long-lost schoolmates up and down Britain can be relied on to complain bitterly that they'd *love* to read your book but can't find it anywhere. No matter how many were supposedly distributed, it is an unwritten law of the book trade that all copies of an author's works are stripped from the shelves whenever said author, or anyone he or she once met in a bus queue, is so much as suspected of loitering in the vicinity. This is done for the humane purpose of protecting writers' egos from over-inflation.

The book industry's parting boot in the groin comes all too soon after publication day, when it becomes clear what your long literary toil has been leading up to: providing the vital raw material for the remainder trade.

Remaindering is simply the flogging off of 'no longer profitable' book stocks at a trade price so low as to break the barometer. I recently had an unusual letter from one publisher. It was a personal note signed by an editor I knew, apologizing for a book's imminent remaindering, telling me how many copies were in stock, and offering me first chance to buy them at the price offered to remainder distributors.

This was astonishing. I was being informed about remaindering *exactly as specified in my contract*! That's not how it's usually done.

The normal routine is for some relative or friend in a far-flung county to mention that for months now, Cheap'n'Nasty Book Bargains just down the road has been selling your novel for peanuts. This seems strange, as your contract definitely says that remaindering won't even be considered for months yet, and that you'll be the first to know.

Authors become cynical about the fact that this unloading happens in remote places, as though to avoid any awkward questions from the actual writer. (One outfit, notorious in the trade, openly offers to dump remainders discreetly on the Continent.) Your editor feigns total bafflement; the evil deed has been done by mysterious 'marketing people' who know nothing of your contract.

The point of informing you is that it gives you a chance to stock up with copies to cushion your retirement – or at least insulate your attic. You could even buy the lot and avoid the stigma of appearing in remainder shops at all. Well, that's the theory....

I tried it. I bought as many pre-remainder copies of Book A as I could afford. They weren't that cheap, but the publisher swore this was – as per contract – the lowest wholesale price. Then I discovered stacks of them at half that price, *retail*, in Oxford Street. Result: after much aggro, a grudging refund.

There was similar trouble with Book B. I bought 'the lot' ... and then found my market spoilt by cheap, dumped copies of the Australian edition which had somehow never been exported. Result: nothing to be done without a lawsuit.

Book C was remaindered early in breach of contract. This sounds illegal, but in practice all the breach does is terminate the contract: by remaindering, the publishers have washed their hands of you anyway, so *they* don't care. 'Can I buy copies?' I asked. 'We only have 20,' they replied, and instantly sent them – with an invoice for some horrendous sum, three times the rate I was quoted by the remainder dealer who'd bought the rest. Result: when shouted at long enough, the publishers decided that the 20 copies had been complimentary ones all along.

Book D was also remaindered in breach of contract. By now I was learning to complain loudly and often, and with the aid of the excellent Society of Authors got an ex-gratia payment in partial compensation for the publishers' crimes.

Book E ... well, the message is becoming clear. When remaindering looms, you need to watch your publishers with laser-eyed scrutiny and be ready to complain until you're blue in the mouth. This is unfortunately not too good for author-editor relations (when, for example, you ask every week, 'Have you remaindered me in breach of contract yet?'). But the price of solvency is eternal vigilance.

Meanwhile, I have this nightmare which starts: 'Hello, this is *8000 Plus*. We have 112,876 Langford columns in our warehouse and need the space. We can offer them to you at 25p each if you take the whole lot....'

8000 Plus 47, August 1990

Dictionary of Quotations

It is well known that everyone who is anyone uses a PCW – I checked by looking back through four years of *8000 Plus*, and the theory was amply confirmed. Not only does everyone use the PCW, they all write about it and all make the same LocoSpell jokes. (Perhaps there should be a five-year moratorium on anecdotes with hysterical punchlines like, 'And for my name, it suggested *Landlord*, ha ha ha ha ha!')

Browsing in other standard works of reference, I idly wondered what bygone pundits of literature and reality have said about their PCWs. Here is a selection from what I found ...

Macbeth's fatal flaw, besides ambition, was that he never mastered BASIC programming commands – the software he sold kept coming back with letters of complaint. *We but teach / Bloody instructions, which being taught, return / To plague their inventor.*

His old mate Banquo, meanwhile, complained that although he had no difficulty getting his PCW to do trifling things like addition, he got into trouble with serious stuff like his VAT return: *And oftentimes, to win us to our harm, / The instruments of darkness tell us truths, / Win us with honest trifles, to betray's / In deepest consequence.*

James Joyce, after whom the machine was of course named, reviewed but in the end decided not to use LocoSpell: *None of your cumpohlstery English here!* Apparently he wasn't too pleased by what the spelling checker made of his more moving passages like: *Finn, again! Take. Bussoftlhee, mememormee! Till thousandsthee. Lps.* (etc.).

Ambrose Bierce of *The Devil's Dictionary* fame had hard words for spelling checkers, too: **Dictionary, n.** *A malevolent literary device for cramping the growth of a language and making it hard and inelastic.* Literary critics suspect that he never mastered the feature which lets you add your own words.

St Paul lost his temper trying to set up baud rates for MAIL232 file transfer: *Evil communications corrupt good manners,* he jotted on a nearby scroll.

That obscure 19th-century poet the Rev. Cornelius Whur liked a nice, clean, shiny screen on his 'Joyce', and recorded this preference in verse: *What lasting joys the man attend / Who has a polish'd female friend!*

But Andrew Marvell found the joys too long-lasting, especially when moving through a big document, and expostulated: *Had we but world enough and time, / This coyness, lady, were no crime ...*

The poet Longfellow had difficulty in upgrading from a low-tech quill pen, and kept hitting the wrong cursor keys: *I shot an arrow into the air, / It fell to earth, I know not where.* (Usually on the wrong page of the file.)

Percy Wyndham Lewis of the Vorticist arts movement tried, not very lyrically, to hymn his favourite software company: *I said (and I always say these things with the same voice) / 'Say it with locomotives ...'*

Byron found the CP/M manual incomprehensible, and cursed its author something rotten: *Explaining metaphysics to the nation – / I wish he would explain his Explanation.*

Indeed, the prophet Jeremiah was even less enamoured of this manual, and suggested extreme measures: *And it shall be, when thou hast made an end of reading this book, that thou shalt bind a stone to it, and cast it into the midst of Euphrates.*

Wittgenstein philosophically added, *Wovon man nich sprechen kann, darüber muss man schweigen*, which translates roughly as, 'I couldn't understand the blasted manual either.'

But A.E. Housman, being a textual scholar as well as a poet, had no sympathy for people who can't follow the instructions, even in Latin: *Three minutes' thought would suffice to find this out, but thought is irksome and three minutes is a long time.*

Jonathan Swift was typically grumpy about the cheapness and ease of PCW word processing, and made some digs at it in one of the satirical bits of *Gulliver's Travels*: *Every one knew how laborious the usual method is of attaining to arts and sciences; whereas by [this] contrivance, the most ignorant person at a reasonable charge, and with a little bodily labour, may write books in philosophy, poetry, law, mathematics and theology, without the least assistance from genius or study.* (I can't imagine how he left out best-selling novels.)

Hugo Gernsback, a much more awful SF writer, tried hard to predict the modern PCW in 1911 but got several small details wrong: *He attached a double leather head-band to his head. At each end of the band was attached a round metal disc that pressed closely on the temples. From each metal disc an insulated wire led to a small square box, the* **Menograph** *... He than pressed a button and a low humming was heard; simultaneously two small bulbs began to glow with a soft green fluorescent light.* You could argue that with this gadget that 'entirely superseded the pen and pencil', Gernsback came within shouting distance of describing glowing green screens and was the first SF writer to predict word processor disks, even if they're in a slightly unlikely place.(By the way, I'm sure I've seen several pictures of the Menograph in Glen Baxter cartoons.)

Thomas J. Watson of IBM was obviously just jealous of Amstrad: *I think there is a world market for about five computers.*

Wordsworth found himself impressed by LocoScript's multiple alphabets but like everyone else became sarcastic about its speed: *Characters of the great Apocalypse, / The types and symbols of Eternity.*

And the forgotten James Grainger, in a 1759 epic poem which must surely have been dedicated to Amstrad (it's called *The Sugar-Cane*), lyrically pinpointed the common factor of all known software: *By microscopic arts, small eggs appear, / Dire fraught with reptile life; alas, too soon / They burst their filmy gaol, and crawl abroad, / Bugs of uncommon shape.*

8000 Plus 48, September 1990

Getting Together

Writing is a lonely business, often made lonelier by family members who don't appreciate that the Creative Process involves hours of staring blankly at a blank screen while scratching your head, sucking your teeth, picking your nose, etc. Talking shop with others who understand this vital fact is half the charm of 'writers' workshops' – our subject of the month.

All those I've been to have a science-fictional flavour. The most demanding is the annual Milford SF Writers' Conference, started in the 1970s as a spin-off from the American event in Milford, Pennsylvania: the resourceful Brits held theirs in Milford-on-Sea, Hants. It's moved around since, and to those in the know, 'Milford' is also a little-known SF spelling of 'Cheltenham' *[or 'Margate', or indeed 'Snowdonia']*.

At Milford, a gaggle of SF writers fills a hapless hotel for a full week. You must have sold professionally (one sale will do), and must bring a stack of copies of your work for discussion: no one is allowed to shred others' stories without offering up their own sacrificial goat.

The daily Milford routine strikes me as a good general model for workshops. Each morning, everyone frantically reads and makes notes on that day's contributions. Each afternoon is a long critical session in which several pieces are discussed into the ground ... after which the conference staggers, pale and sweating, to the bar.

The system is designed to give everyone, no matter how shy, his or her say on every manuscript. Someone is chosen to start off (by drawing straws, by asking who actually wants first go at this MS, or by dictatorial decision of whoever's in the chair). Everyone in turn has a nominal three minutes to comment – it might be a hymn of unsullied adoration, a devastating attack, or detailed DIY instructions for dismantling the story and putting it together so it'll fly better and further.

The victim in the hot seat can then reply to all these tormentors at any reasonable length. (It was at this point that one very famous SF author reputedly cried, 'You bastards, how dare you find fault with Me?' – and stalked out, never to be seen at a Milford again.) A final free-for-all discussion, a five-minute break, and the next MS goes under the microscope.

On one or two evenings, there will probably be a red-hot debate on some topic which, like erupting magma, overflowed the bounds of the timed afternoon sessions.

On the other evenings ... I think we'd better draw a veil over the libellous shoptalk, in-jokes and wildly silly literary games. It's always a great week, though always too expensive. The groan-laden morning after the final party is punctuated by eldritch screams as participants receive their hotel bar bills.

Much cheaper and much more frequent are the writers' gatherings confined to one day or weekend at someone's house. The discussion sessions tend to have much the same structure as at Milford; what varies is the reading of the actual manuscripts.

If you are well organized, they're circulated in advance by each contributor or by the current meeting's host. This host needs to send out reminders of everyone's address, and to exercise fascist control over the number of actual contributors – unless the attendance is tiny, your brains will fall out should you try to give more than four or five stories the full treatment in one day.

Less efficient groups can ask people to bring many copies of their MSs to be read on the spot. In a one-day session, this always means hasty skimming by latecomers, and only the ablest critics will be able to muster more than 'quickie' first reactions. There's a disreputable school of thought which says, 'Who cares? It's the social side I come for.'

Totally disorganized groups, like the first and most enjoyable one I ever belonged to, madly rely on the stories being read aloud to the workshop members and commented on after a period of mature reflection lasting about five minutes. One aggrieved author said, 'Langford, I could *see* you counting the pile of unread pages and openly calculating how much longer you had to suffer....' Others read their stuff so well that criticism of the actual words, however lacklustre, is lost in admiration of Performance Art. Totally unfair, of course.

(Still, several regulars from this series of chaotic meetings later became SF household names. You'll all have heard of L. Ron Hoover, Arthur C. Kellogg and Isaac Amstrad.)

Even more economically, far-flung groups can conduct these critical sessions in slow motion, by post. One popular system has a packet of MSs circulating as a round-robin parcel, each recipient passing it on with *(a)* critical notes on the contents; *(b)* a new contribution. When the package comes round again, you remove your by now tatty and coffee-stained MS, and read the huge wad of criticism which has accumulated.

Under the generic name Orbiter, several groups like this are run in connection with the British SF Association *[here I have the current membership secretary's address, but now see www.bsfa.co.uk]*.. Or you could simply locate a few other aspiring authors yourself.

Of course, if you know other PCW-owning writers, a lot of postage can be saved on the round robin. Several stories and a lot of critical response can be packed into even the 360k of a single, circulating 8256 disk. Use a lightweight (reusable) jiffy bag and have the writers add their thoughts in turn to the end of each story document – more economical with disk space than extra files of comment.

Many of you, I know, are doing it already. Good luck, and watch out for those 'traditional narrative elements' ... the delicate Milford euphemism for clichés.

8000 Plus 49, October 1990

Half Century

Fifty issues of *8000 Plus*, and I've appeared in them all, sometimes twice. It's like being the Oldest Inhabitant, with a long white beard constantly getting entangled in the space bar, and several address marks missing from my forebrain. Pausing only to open the congratulatory magnum of champagne which Future Publishing did not send, I poked through my dusty archives of the magazine's prehistory, stretching back through eras of quill pens and clay tablets to the beginning of all things in, actually, 1986.

How embarrassing. Here's the yellowed printout of that first column, when I had no idea how the magazine would look or what my page should be called. Having recently read the unfunniest funny book in literary history, I pinched its title from sheer desperation: *Diary of a PCW Nobody*.

(Incidentally, copyright law turns a blind eye here: I keep being fooled by new books which recycle other authors' titles, but no one ever complains. Thus *Double, Double* is a 1950 whodunit by Ellery Queen, and a 1969 SF novel by John Brunner – both quoting a 1606 hit by Bill Shakespeare.)

Luckily my first title never reached print: editor Ben Taylor (or maybe Simon Williams, with whom Ben kept swapping in the first year) schizophrenically changed it to *Langford's Diary* on 'Opening Menu' and *Langford's Printout* within. It stayed as the latter until Rob Ainsley grabbed the editorial hot-seat at issue 20, and presumably thought long and hard about the fact that no printout was involved: I was sending in the stuff on disk.

So, as of issue 21, I've lurked under the scarcely modest or self-effacing column title *Langford*. This is none of my doing. I am shy and retiring, but as Tolkien wrote, 'Do not meddle in the affairs of Editors, for they are subtle and quick to anger.'

Indeed, Ben once became subtly miffed when I included some rude cracks about overpricing of small-press books – little knowing his ability to vanish into a phone booth and assume his alter-ego as a director of Kerosina Books, whose productions were (and are) very nice but not all that underpriced.

Next in my cobwebbed files comes a stack of letters from Rob Ainsley, some very strange indeed: 'A mate of mine has named his house 'Freepost' and swears he knows a Swedish secretary called Per Pro.'

My favourite is Rob's list of editorial woes. 'Number one is the old LocoSpell article, cataloguing its bizarre replacements in a wide-eyed first person narrative. Second is probably Case in Point, invariably restricted to LocoScript. Extra naff points are notched up for pet PCW names ('I have a new girlfriend. She's called Joyce. My wife wonders who she can be,' etc.), wally club activities ('My PCW is invaluable in the running of my Civil War Re-Enactment Society. I keep a costume list on file ...') and daft mistakes ('I use Logoscript exclusively ...' 'I am a regular reader of *Amstrad PCW Plus* ...').'

Along came issue 31 and spanking-new editor Steve Patient, whose total commitment to Amstrad technology was shown by the fact that all his notes to me were scribbled in smeary pencil on both sides of Future Publishing compliments slips. 'We are very bored with the bad cartoon of you which fails to decorate your column,' he grumbled in early 1989, and those strange pictures of me up there have become stranger ever since.

Steve liked to cheer the hearts of depressed authors whenever possible: 'Dear whingeing sod, We're paid to put gratuitous words on paper, and have a quota to meet. Those inserted into your text were just a few I had left over.... Are we still paying you? We are? Who wants to read about *statistics* for Ishtar's sake? I wouldn't read it and I have to.' *[This reference would seem to be to the May 1989 column – see page 68.]*

Or, when I suggested doing an article on assembler and, specifically, an alternative SUBMIT.COM I'd written which could be used on write-protected start of day disks.... 'This sounds like a thoroughly useful piece of programming – pity I can't see any way to use it.'

It was Steve who published my super-pedantic column on How To Punctuate Real Good (page 39), which produced more correspondence than any other. A friendly ex-newspaperman wrote in to explain the original reason why tabloid papers break stories into the shortest possible sentences and paragraphs: in hot-metal typesetting, cutting to fit was most easily done by removing whole paragraphs of type. So they had to be short.

Or to put it another way, that breathless style with a new paragraph for every sentence stems from a now bygone technology, even though all today's major papers are electronically edited on DTP systems which can reformat stories with the utmost ease. I might have guessed.

At around this time I put out some feelers about titivating my *8000 Plus* columns for collected book publication.

'They are brilliant, but too ephemeral and magazine-ish,' replied a typical publisher.

'But when I've rewritten them, they will be not only book-ish but even more brilliant,' I modestly pointed out.

'Perhaps, but all the PCW owners will have read them in *8000 Plus* and no one else will be interested.'

'Um, I could rewrite them further to make them of wide general interest ...'

'In that case the book would not have any identifiable market at all.'

The message is clear. If you like these columns, don't throw away your old copies. Also, you are not an identifiable market. Sorry about that.

8000 Plus 50, November 1990

• *Happily Cosmos had no objection to this much delayed book, or colonnade.*

Legal Frictions

Freelancing as a writer can involve you with sleazy operators who sooner or later may display the cloven hoof. Earlier this year I found myself suing publishers in the Clerkenwell County Court – all by remote control, without having to walk further than a nearby letter-box. You might need to do the same one day, so here's the terrible saga.

Before, I'd always been lucky: my publishers paid what they owed me, eventually. (Well, there was one dear old chap who sold his firm and retired: he reckoned the new owners inherited all debts; the new owners reckoned not; in the end I wrote off the outstanding four quid.) This luck changed with the final instalments of my SF review column for the now defunct games magazine *GM*, published by Croftward Ltd – accursed be their name.

Learning that even the editors were no longer getting paid, I began to worry. Cheques continued not to arrive. Panic! This was where the SF grapevine came in handy: I was soon in touch with a writer (Marcus Rowland) who'd already successfully sued this outfit. Picking his brains eased my first foray into the deadly arena of Small Claims litigation....

Most of us dread legal action, feeling with some justification that going to court is equivalent to lighting a huge bonfire of banknotes around which solicitors and barristers will dance a merry jig for exactly as long as your life savings last. But mere writers tend to be chasing relatively small sums, and if the amount is under £500 there's no question of solicitor's fees.

So ... suppose you've been owed money too long by some rotten, lousy publishing firm which out of cowardice we'll call Ripoff Ltd. 'Proof of debt' is the basic requirement: a contract or acceptance letter which mentions a fee should do, together with a copy of the book/magazine/newspaper containing your work. If you have only the latter, some research might be needed to establish their standard payment terms. ('Payment on lawsuit only.')

You need to give advance warning in a formal letter. Typically you'll have sent half a dozen and got no reply, but a final one is still necessary – containing the magic words, 'Unless I hear from you within seven days I will issue proceedings in the XXXX County Court ...'

My own inclination is to *say* 'seven days' but actually allow more leeway for a dreamy interval of Post Office limbo. However, I might have done better to rush as fast as possible ... as will emerge. Meanwhile, there's the question of which county court to name: for defaulting publishers, it's the one in the district containing their Registered Office. Don't worry if a check at Companies House shows Ripoff Ltd to be registered in the Orkneys: you needn't visit.

Now it's time for some official forms, available free from your own local County (not magistrates') Court. You want a 'Request for Issue of Default Summons' and the accompanying Form EX50, actually a lucid little booklet about the whole procedure. This contains an easily lost slip of paper listing court fees – currently 10% of the debt you're chasing, with a minimum of £7 and a £37 plateau for debts of £300 to £500. Get two copies of the summons form, my pal advises – you might spoil the first, although I smugly didn't.

While you're there, pick up forms EX50B and C, explaining how to translate a court victory into actual debt recovery, and how much *this* will cost. Nobody told me about that bit until rather too late. *[Some extra plain-English booklets have since been added.]*

By the way, the useful EX50 booklet contains examples of how warning letters should be phrased, and specimen Statement of Plaint drafts. Perhaps some public-spirited soul will one day put together a public domain disk of such documents and information: the PCW Small Claims Kit, with all the standard forms of plaint ready to load into LocoScript for adaptation to your own case.

'Plaint' is mere legalese for complaint, and after a heading which names the court, leaves room for a case number and gives both parties' addresses, the writer's Statement would consist roughly of: 'we agreed this; they published that; payment not received by whenever; letters requesting payment sent then, then and then; letter warning of court action sent on such-and-such a date; the Plaintiff claims £XXX, being the fee owed.'

There's a box measuring four by five inches for all this on the summons form, which is why it's a good wheeze to attach a separate and perfectly word-processed statement instead. (This, indeed, is how my pal ruined his first form – trying to get everything into the box.) In fact you should print and attach two identical copies.

You then post everything to the relevant court – form, statements, court fee, and the traditional stamped addressed envelope in which you'll be sent an acknowledgment carrying the all-important case number, to be quoted *ad nauseam* in further letters.

It went like clockwork. The court served a summons on my loathed publishers; two weeks later, as explained in the booklet, I bunged in the 'Request for Entry of Judgement by Default' form (also free from the local court), and was awarded the full amount of my debt plus the £37 fee.

All I had to do was collect it.

I puzzled at length over booklet EX50C, 'Enforcing Money Judgements in the County Court'. Sending in bailiffs to seize the company's battered PCWs would cost me £38 and might not actually pay off, since the stuff would be sold by public auction for peanuts (how can I get to these sales?). Around then, the wretched company announced its creditors' meeting, and the first thing that emerged from the accounts was that all their worldly goods were forfeit to the bank as security on the *third of a million* owed them....

Paradoxically, a company that's a million pounds in the hole can laugh at small creditors: the big ones (the bank, VAT, Inland Revenue, etc.) are now calling the tune and couldn't care less about ordinary bruised people.

Your hero emerged £37 poorer, but with a wealth of exciting new experience and a stern moral. The moral is: don't start even a small-claims action against vile defaulting publishers, until you're sure there are some assets for you to grab.

8000 Plus 51, December 1990

Hack's Quest

Welcome to *Hack's Quest*, the interactive game that makes the *Hitch-Hiker* adventure look like a Conservative Party conference, and *Nightmare on Elm Street 27* look like a repeat. In this sizzling intellectual challenge you play a freelance writer with an article to deliver. The editor has just moved your deadline two weeks forward because of an unforeseen cataclysm called the New Year. It is a bleak Monday. You have a hangover.

Type your commands at the > prompt. (Enter HELP to ask for assistance.)

You are in an indescribably sordid hallway. Shabbily carpeted stairs lead up to your workroom.
> HELP
Kindly remember you are a freelance writer. That is: you're on your own, sunshine.
> GO NORTH
Stop kidding around. You have no idea which way is north.
> GO UP
That's better.... You are in a grimy workroom, whose stale, dead air is thick with the sweat of old deadlines. Bills litter the few square feet of available floor. Like an altar-idol dominating the foul temple of some obscene cult, the PCW broods over the shambles.
> TURN ON PCW
You turn savagely on the computer, but it is totally unmoved by your threatening attitude.
> SWITCH ON PCW
You can't do that.
> WHY NOT?
In your eagerness to rush to the pub last night, you forgot to switch the machine off. It now seems very warm. An electricity bill is prominent in the heap at your feet.
> INSERT WORK-IN-PROGRESS DISK
I see no disk.
> SEARCH FOR DISK
An hour passes. Eventually you find the disk under some rejection slips in the far corner of the room.
> INSERT DISK

Are you sure?

> ER ... WIPE DISK WITH HANKY AND INSERT

By first cleaning off about an ounce of revolting tangled hairs and dust-bunnies glued by static to the disk, you have avoided the immediate, catastrophic failure of your drive. (Score 1 point.)

A blank screen confronts you!

> WRITE ARTICLE

You can't do that. You have no inspiration.

> SEEK INSPIRATION

Where do you want to look? There is a book here. There are bills and rejection slips here. There are old copies of *8000 Plus* here. There is a bottle here labelled 'Inspiration: Matured 10 Years in Oak Casks at Glengrotty', but it is empty. This is why you have a hangover.

> EXAMINE BOOK

The book is *Roget's Thesaurus*. You study it for inspiration and find excitation, possession, afflatus, exhilaration, intoxication, headiness, encouragement, animation, incitement, provocation, irritation ... As usual, the irritating truth is that there's no inspiration to be had from Roget. But one keeps hoping.

> EXAMINE MAGAZINES

They are full of good stuff. You wonder whether you could steal something from older ones which everyone must have forgotten.

> YES! STEAL LOTS OF IDEAS

Unfortunately your rigid moral code prevents you from actually doing this.

> OH

But as you stare blankly into space, the shadow of a notion begins to take shape in your mind.... The doorbell rings!

> IGNORE DOORBELL

It might be someone with a cheque, or Steven Spielberg's office boy asking after movie rights to your articles on How To Write Assembler Real Good.

> GO TO DOOR

You are in the sordid hall. It was the postman. Why did he ring? Peering blearily at the mat, you find a card saying: 'You failed to answer the door within the prescribed 0.5 nanoseconds, and three valuable-looking parcels have therefore been rushed back to our depot two hours' walk away.' The rest of the mail is all brown envelopes with sinister windows in them.

> GO UPSTAIRS

You are in the grimy workroom. Your train of thought, such as it was, has been completely derailed and lies upside down next to the tracks.

> SIT AT PCW AND WAIT

As you stare blankly into space, the shadow of a notion begins ... The telephone rings!

> IGNORE PHONE

Are you kidding? You deliberately chose one whose tone rattles the windows of houses across the street, to make sure you never miss an important call from your publishers. Or Steven Spielberg, of course.

> RELUCTANTLY ANSWER PHONE

A hollow voice says: 'Good morning! Have you thought how much you could enhance the value of your crummy home by ripping out the windows and installing expensive double glazing, covering up the original Victorian brickwork with synthetic cladding in a lurid shade of pink, and replacing that out-of-date slate roof with fibreglass simulated thatch?'

> SAY NO AND SLAM PHONE DOWN WITH UNNECESSARY VIOLENCE

Ok.

> WAIT

As you stare blankly ...

> TAKE PHONE OFF HOOK

... you are inspired with a sudden, blazing need for coffee!

> WHAT?

Coffee. A beverage made by brewing the roasted and ground seeds or beans of a tropical evergreen shrub 8 to 10 metres high belonging to the genus *Coffea*, of the Rubiaceae, or madder family. Or in your case, instant from the Co-op.

> SIGH. MAKE COFFEE

You have wasted valuable time catering to your selfish wish for coffee. It is only two hours to the last postal collection! The article must be finished and printed out by then!

> HELP

We have already been into this. Remember, you pay Class 4 National Insurance contributions, which bring you absolutely no benefits but are purely and simply an extra tax levied on you for your temerity in being a self-employed person. You are, in other words, the scum of the earth. You expect help?

> PLEASE

You have discovered the magic word! (Score 1 point.) Here is a hint. To solve this puzzle you must find and read something you have not yet studied.

> EXAMINE WHISKY BOTTLE

The small print says that it contained known carcinogen E6234, permissible colouring, fusel oil and monosodium glutamate.

> EXAMINE PCW

There is a message on the keyboard! It begins: 'TAB QWERTYUIOP'.

> TA VERY MUCH. EXAMINE MAIL

The first envelope contains a bank statement. It is horrifying! Your overdraft exceeds the poll tax deficit of many small boroughs. Staring at it in terror, you feel all your moral inhibitions dissolving. (Score 1 point.)

> AHA! STEAL IDEAS FROM OLD MAGS AND WRITE ARTICLE

Ok. You have now shed your scruples and demonstrated the

qualities required to survive in 1991 Britain. (Score 2 points.) You have completed Level 1 of Hack's Quest.

Your score was 5 points. You have graduated from No-Hoper and now qualify as a Grubby Hack. In Level 2 you will confront the thrill-packed challenges of the Plagiarism Suit, the Cirrhosis Clinic and Writing A Novel.

Continue now?
> QUIT GAME
You can't do that.

8000 Plus 52, January 1991

• *This was the column which came out looking most surreal on the page. Apparently, when the text was moved to a Macintosh for page layout before printing, the software flipped at the 'prompt' signs. As a result, all paragraphs starting with > were omitted.*

Tweedledum and Tweedledos

One of the most tiresome things that computer owners do is to play the ever-popular game 'My Computer's Better Than Yours, So There.' You know: my PCW is better than your IBM because it fits more text on the screen, but my IBM is better than your PCW because it's 'state of the art' (i.e. costs more).

Most boring of all are operating-system snobs who drone on about the superior virtue of CP/M or MS-DOS – whichever they happen to use. This one-upmanship obscures the interesting fact that if you know the basics of CP/M, you automatically know a fair bit about DOS as well ... and vice-versa. Forget the tedious arguments and revel in the knowledge that when confronted with an IBM (which ghastly fate can happen to any of us), you can suss a great deal by remembering elementary CP/M.

If you prefer to ignore the whole issue and use LocoScript for all 'file housekeeping' operations, you have my sympathy, not to mention permission to skip what follows.

In both operating systems, the basic 'type something and then press Return' prompt looks like this: A> – carrying the additional message that Drive A is being used. You can change it by entering **B:** to select drive B. IBMs often have a hard disk C instead. Challenge to the intellect: what does the command **C:** do? Or on the PCW, how about **M:**?

(I'm not going to keep mentioning this, but each of these CP/M and DOS commands takes effect when you press Return or Enter following the command.)

Typing **DIR** displays the directory, although its appearance differs in DOS. (The DOS command **DIR /W**, where /W stands for 'Wide format', gives a more CP/M-like look.) Another familiar command, **TYPE FILENAME.DOC**, shows the contents of a 'plain ASCII' text file on the screen. CP/M does better here, halting at each full screen with a friendly 'Press Return to Continue' ... DOS demands either hair-trigger reflexes on the Pause key or, again, a more complicated command.

Deleting files is 'sort of the same'. The CP/M command is **ERASE** and DOS prefers **DEL**; but as a concession to CP/M visitors, DOS will in fact accept **ERASE**;

however, in CP/M we usually abbreviate it to **ERA**, which DOS refuses to recognize. **ERASE *.*** remains deadly anywhere.

Certain commands use the same keyword but work 'backwards' in DOS. ('You mean they work backwards in CP/M, fool,' writes Outraged DOSser of Tunbridge Wells.) **REN** or **RENAME** is the obvious one. When I want to pass off an old column on a new editor, I rename it in CP/M with **REN COLUMN.NEW=COLUMN.OLD**. On the DOS machine, the same effect requires **REN COLUMN.OLD COLUMN.NEW**. Challenge to students of relativity: which one is backwards? Perhaps DOS makes more sense here; there's something very *computer-programmerish* about the CP/M command.

I've heard it said that CP/M is better than DOS because many useful commands are built in and don't need program files. I've also been loudly told that DOS is better for the exact same reason. An explanation is required here.

Take **REN**: in each case, it both is and isn't 'built in'. CP/M does have **REN** built in, since it will all on its own accept a correctly typed **REN THAT=THIS** command. But if you mistype the THIS filename as something nonexistent, the program file RENAME.COM is needed in order to issue the terse error message 'No File'. Similarly, **DIR** can always be used but needs DIR.COM for all the fancier options.

In DOS, the most-used commands live in a file called COMMAND.COM which is always loaded ... so they're all built in, *except* that depending on your set-up COMMAND.COM tends to get overwritten when other programs run, whereupon DOS insists on reloading it from disk immediately. You need it around.

(Hardened operating system users on both sides have much the same solution to this recurring need for a .COM file or files: their 'start of day' disk copies needed CP/M utilities, or COMMAND.COM, to a memory drive like M:, and uses special commands to tell their operating system where to look for the needed files. Of course if you have a hard disk, which all commercial IBMs now do, COMMAND.COM lives there and reloads imperceptibly.)

The biggest bone of contention is **PIP**. Do you prefer to type **PIP NEWFILE=B:ORIGINAL** or the DOS equivalent **COPY B:ORIGINAL NEWFILE**? DOS fanatics sneer that PIP isn't 'built in' even rudimentarily: you always need PIP.COM. CP/M acolytes retort with boasts about the million extra optional things that PIP can do to files while copying them.

Let us rise above sordid bickering, and merely note that this is one difference you do need to remember. Typing PIP in DOS results in the greeting 'Bad command or file name', while CP/M reacts to COPY with incredulity: **COPY?** There's one similarity, though – the 'verify' option which requests that extra care be taken to make an accurate copy. PIP with [V] at the end of the command corresponds to COPY with /V.

Last oddments.... In both systems, files ending in .COM are programs which can be run by typing the filename without the .COM. (To make life more interesting DOS has another flavour of program file that ends in .EXE but is used in exactly the same way.)

CP/M lets you create SUBmit files containing several preset commands: if MYSTUFF.SUB contains your favourite commands, you can issue the whole lot in sequence with the aid of SUBMIT.COM, by typing **SUBMIT MYSTUFF**. Again DOS is eerily similar, with its corresponding BATch files containing 'batches' of command lines. You could copy MYSTUFF.SUB directly to MYSTUFF.BAT and use it in DOS by typing **MYSTUFF** (no SUBMIT.COM required) ... although not all the commands in it might actually, as it were, work.

And both systems come with appalling 'text editors', ED and EDLIN respectively, which are almost identically unusable!

Here I planned a theological analysis of the parallels between CP/M's division of disks into 'groups' and DOS's organization into 'subdirectories', but I see that men in white coats are beginning to hover round. In the 1970s, someone published a massive tome which 'proved' by linguistic analysis that the Hebrew language was the same as Greek. Let's not go that far – but knowing your way around CP/M remains a good basis for coping with the mysterious and pervasive terrors of DOS.

8000 Plus 53, February 1991

Index, Peculiar

Many years ago I completed my first full-length book and retyped the whole horrible thing in a fair copy ... ah, those ghastly days when peasants tilled fields with pointed sticks, doctors clamped leeches on your tender parts and authors bashed out typewritten drafts. Then it was time to prepare the index, another gruelling task.

The index sorting system was marginally high-tech, in that the entries were handwritten on Fortran computer cards provided by a generous employer (the Ministry of Defence). My reward for this drudgery was that the editor who bought that book has since published lots himself, each with a bibliography which cites *War in 2080: the Future of Military Technology* by D. Langford, whether or not it's relevant. Like most of the books he lists, it invariably has a star against it; as editor/author John Grant explains, 'I have indicated by * a book which has a lousy index.'

Thanks, boss.

Yes, indexing is an art, even when your computer shoulders the burden of actual sorting. Since those laborious days I've written various indexing programs for use with word processors (AnsibleIndex for LocoScript is the one I am not going to plug here), and still find there are many creative human choices to be made at both ends of the process. First you pick which words, phrases and themes are to be indexed: for example, it's amateurish to include mere 'passing references' like that to the Fortran programming language above. When the program has done its sinister work you'll want to edit and titivate the text.

It's worth taking expert advice. One good guide is *Indexing, The Art Of* by G. Norman Knight (1983). Also, the British Standards Institution does several pamphlets on the niceties, like BS 1749:1985, the definitive word on alphabetical arrangement ... which all by itself is trickier than you think.

The Society of Indexers can be contacted at 16 Green Road, Birchington, Kent, CT7 9JZ.

So much for the morally worthy bit. Because indexing is hard work, there's a strong temptation to conceal some little joke in the forbidding columns of the index. These can be great fun to spot.

As an example of something witty rather than actually funny, a friend of mine wrote a book of which one page contains a tiny puzzle about the author of a long-forgotten work. The mystery writer isn't mentioned in the text, but any reader guessing the correct name will find it indexed in its proper alphabetical place.

Much funnier is the index of *The Clothes Have No Emperor*, a chronicle of Ronald Reagan's presidency by Paul Slansky (1989). The author must have giggled insanely as he compiled the long, long entry on Reagan himself, with something like 140 subheadings:

'Blames Carter ... blames the media ... blames miscellaneous others ... Bond, James, honoured by ... books about ... bullet in chest temporarily unnoticed by ... campaign oratory of ... cancerous pimple called "friend" by ...'

There are 18 page references under 'challenge to accuracy of', 19 under 'inability to answer questions of', 17 under 'macho bluster of', 22 under 'misidentification problems of' and 33 under 'misstatements by'.

It's vaguely reminiscent of the non-story in J.G. Ballard's *[then]* most recent collection *War Fever*, which consists entirely of an index along these alarmingly suggestive lines to a book which now need never be written.

Perhaps the shortest, sharpest and rudest use of the index as a weapon is in Bernard Levin's *The Pendulum Years* (1970), whose account of 1960s Britain naturally includes much about the *Lady Chatterley's Lover* censorship trial – where all the juiciest four-letter words were solemnly bandied in court. If you go to the index and look up a certain reprehensibly female-anatomical term, you'll find it referenced: 'see Griffith-Jones, Mervyn'. He was the prosecuting counsel. Ouch.

From time to time someone has the bright idea of giving a humorous book an elaborate, wordy index full of jolly laughs. Purists are sniffy about this, and the fun is usually poisoned by consumer resistance to sitting and reading through an index. Once in a while, though, you do find something worthy of chuckles.

A.P. Herbert's various books of 'Misleading Cases' have highly tendentious indexes, full of little digs in the ribs. Once, though himself an MP, Herbert chafed at the unfairness of Parliament's failure to observe things like the licensing laws which it imposed on the rest of the country ... so, having tested this convenient immunity in court, he made a list of all the activities Members could presumably get away with in their privileged House and strewed these through the index of *Uncommon Law* (1935):

'ADULTERATED FOOD: May be sold at the House of Commons ... ARSON: Is lawful, in the House of Commons ... BRANDY: May be sold at tea-time, in the House of Commons ... BURGLARY: In the House of Commons, is lawful ... CHILDREN: Born in the House of Commons, need not be registered ... COCAINE: May be sold at the House of Commons ... DIRTY BOOKS: May be sold at the House of Commons ...'

Another of Herbert's books, *What a Word!* (1935), crusaded for better prose with splendid if not very functional index headings like 'Bacilliferous Beverages', 'Cannibal English' and 'Septic Verbs'.

The index as prose poem is illustrated in that famous collection of bad verse *The Stuffed Owl*, edited by D.B. Wyndham Lewis and Charles Lee (1930). This faithfully chronicles all the poets' daftest metaphors, so you find entries like 'Worm, lisping ... militant ... far-fetched, *see* Silk-worm' and can follow the cross-reference to: 'Silk-worm, Spartan tastes of ... sinks into hopeless grave.' There is social comment here too: 'Frenchmen, fraudful, mix sand with sugar' and 'Bilious attack, poetical description of.'

But perhaps the least-known, the most useless and the most grimly instructive index of them all (for aspiring writers, at least) is the one appearing in Hilaire Belloc's satirical *Caliban's Guide to Letters* (1943). Slowly it dawns that every single reference is to –

'Action, Combination of, with Plot, Powerful Effect of in Modern Novels, see Pulping, p.187. Advertisement, Folly and Waste of, see Pulping, p.187. Affection, Immoderate, for our own Work, Cure of, see Pulping, p.187. Amusements of Printers and Publishers, see Pulping, p.187. Art, Literary, Ultimate End of ...'

8000 Plus 54, March 1991

Laser Duel

I used to know a home computer pundit who swore by the Sinclair ZX81. This, according to him, was the trailblazer machine, the one that began it all. Yes indeed, it used the same microprocessor as the present-day PCW, but there were a few little deficiencies. My pal was undaunted.

All right, the horrible plastic membrane keys were no good. He'd added a real keyboard. Also, the picture output to the TV screen was terrible. He'd added a real monitor. And of course the ZX81's data storage on slow, slow cassettes was something of a joke, so he had this huge adaptor box sticking out of the computer, interfacing to a hard disk drive.

That was a few years back, and I like to imagine his desktop today: a seething spaghetti of wires and ribbon cables entirely concealing Sinclair's original plastic box, trailing off to CD-ROM units and modems and, of course, the status symbol for home users ... a laser printer.

This all came to mind when I went mad and bought my own laser printer. It seems perverse to add a printer costing substantially more than the PCW, but there are other computers here too, and anyway I was feeling gloomy and wanted a new toy (too many computer peripherals are bought for no better reason than this).

In 1989, incidentally, the prices of laser printers began to slip visibly. They used to cluster around the £1500 mark, but there's now a fair choice at well below a thousand. You have to shop around with steely eyes, because the Recommended Retail Price con is in operation: 'RRP is £1399 but we can offer it for an astonishing £800!' Just as with hi-fi systems, research invariably discloses that *(a)* nobody anywhere is asking the full RRP; *(b)* everybody else, especially mail-order outfits, is undercutting the first price you heard.

Me? I bought an Epson GQ-5000 by mail order for less than half the official fantasy price, and started connecting it to all the computers in turn. Would it work with a PCW? Oh, the agonizing suspense.

I must confess to imitating one bad habit of real software professionals – rushing madly into things without looking at the instructions. ('Manual? We don' need no steenkin' manual!') My first discovery was that although I could *see* the alternate printer drivers called D630.PRI and FX80_NLQ.PRI on LocoScript 2's disk manager screen, the menu revealed by **f6** 'Settings' was adamant that no printers but MATRIX could be selected. H'mm. I capitulated, peeped into the manual, and found nothing.

A bit of head-scratching solved this one. As supplied, the printer files were in

Group 1. LocoScript was not happy unless you started with them in Group 0, so it could then copy them to Group M for use. I did the copying, went through the selection routine for FX80_NLQ again, and was told 'Printer absent'. Aha: LocoScript contacts the printer with a little chug when you load the start-of-day disk. Time to restart with Shift-Extra-Exit. No, on second thoughts: time to save the new printer setting (same **f6** menu) and then restart.

I'd already connected the hardware. Our 8256 has a CPS8256 interface box on the back, mostly used for serial communication with IBM computers. A standard ('Centronics') parallel cable joined the box to the printer, which can be set up via little buttons on the front to imitate various other machines, including Epson FX models ... hence the choice above. Excitement mounted.

Well, it sort of worked. A beautifully printed sheet slid out. The justification and pitch changes were all right (the printer's basic Courier 10 font looked a bit crowded in 12-pitch, but I could fix that by selecting another typeface with the printer console buttons). Italics, boldface, underlining, super- and subscripted text, all came out perfectly.

Not so successful were the various unusual characters I'd put in the test document. If they were in the standard international character sets, they printed exquisitely: Continental accents like the acute **é**, for example. If not – meaning that LocoScript constructs them as graphics rather than just sending an ASCII code to the printer – they printed as spaces. This was the fate of my Greek and Old English test characters, and likewise of the copyright sign (which, maddeningly, *is* in the printer's character sets).

It seemed a hopeful start. I fiddled with the document and tried printing again. Instantly, nothing happened!

No doubt anyone from Locomotive who reads this will send a strong letter about how I should have bought their disk of printer drivers before even considering this idle experiment. All the same.... Although the laser gadget imitates an FX matrix printer pretty well, there is a failure of communication somewhere between it and LocoScript. After one page, the software gets all petulant and insists that the printer is 'waiting for paper'. Meanwhile, a gigantic stack of nice clean A4 paper sits in the printer tray, ready to feed automatically.

Next came the engineering ritual of 'Keep changing the parts until it works.' I swapped cables: no luck. I changed the PCW for a 9512; the procedure for setting up the printer was much the same, except that over there the dot-matrix driver file is called DMP.PRI. Also, the 9512 meant another cable change: instead of the 'Centronics at both ends' connector previously used, you need a standard *IBM* parallel printer cable.

I'm still looking at the 9512 disk manager screen, which after one print-out is stuck saying 'Paper please' (a phrase which makes me retort 'Comma, please!').

This is the state of the art in PCW laser printing research at the crumbling, poverty-stricken HQ of Ansible Information. Will I grit my teeth, ignore the anguished protests of Barclaycard, and buy more printer drivers? Will LocoScript then continue its mocking claim of 'Waiting for paper'? Will our editor tell me to shut up on the subject? Stay tuned, or not, for another sleep-inducing episode....

[Locomotive wrote to explain with a certain smug pride that the program won't work with any printer but the supplied one unless you invest in their additional drivers.]

8000 Plus 55, April 1991

Bye-Bye BT

This year our company decided to save pots of money by abandoning its 'sophisticated' electronic communications system. For years we'd been struggling with one of the world's worst, a nationwide network seemingly designed by a computer hacker who once went to business school but failed. You want me to name names? I will name names. Telecom Gold.

The idea is simple. You have a computer, an interface box and a modem. You ring up the system to send or receive text messages. (You can also download information from databases and suchlike.) The PCW's MAIL232 program can just about cope with this, although for regular use you'd want some software that automates the typing of fiddly passwords.

One thing you soon discover is that few people use Telecom Gold. If they're poor they write letters or make phone calls; otherwise they use telex or fax. Fax costs more and is technically inferior: messages arrive as printed images, and if you want them on disk they have to be typed in again. This is where TG should score, but it's such a pain to use that few persevere.

Taking it blow by blow ...

On joining you receive a confusingly arranged booklet with a lousy index and several errors. This insists that TG is easy to use ... but the information on setting up those complicated communications things like data and parity and stop bits (without which you can do nothing) is buried on page 58 in a chapter mystifyingly titled 'Off-line preparation'. The setting-up process is called 'configuring' by TG: there's no index entry in the booklet for this or any plausible synonym.

Such disinformation is apparently a ploy to extract more money. A fatter manual is available, for a fat price. It's as though your phone subscription only entitled you to edited highlights of the directory.

You dial up the TG computer. If you live outside London you'll want to save money by connecting via a 'PSS' number at local call rates (though many places are so remote and barbaric that this isn't available. Wiltshire, for example. More expense). The TG booklet lists PSS numbers, but doesn't mention the secret extra characters you must type before PSS will accept the connection. After that you simply type a 13-character code, followed by a 10-character PSS identifier, after which Telecom Gold requires your 9-character customer ID, and a password ...

When you finally get all these countersigns right, you are in TG itself! Often this doesn't work first time, or the system disconnects you mysteriously, or is out of action for 'necessary maintenance' ... we have a theory that BT bought a second-hand computer from NASA or somebody, which periodically halts while men in overalls prod it with screwdrivers. A backup machine? That would cost money. Meanwhile, you are *personally allowed* to pay for all those wasted phone calls.

At last the glories of Telecom Gold await you.

TG uses plain text only. To clear the screen it laboriously sends 25 carriage returns and line feeds. Posh layout consists of long rows of spaces, hyphens or asterisks chugging along the line. Occasional lines are too long, so that text wraps amateurishly in mid-word. Punctuation and spelling are erratic. You are, incidentally, charged for every character you receive.

TG does not have any system for correcting transmission noise (which is of course *totally unheard-of* on BT lines): so often a bit of text will l{{h;q li}kw th}}/?s. I *said,* a bit of t{{!qj will look like this. You pay if you want the garbled bits repeated ... per character and per second of connection time.

TG is a business system, so naturally you'd expect to use currency symbols. Yes, but only dollar signs. If you send the message 'You owe me £100', the £ vanishes ... same with the yen sign and every single accented character in the ASCII set. BT chose an antiquated transmission setup which handles the text characters 32 (space) to 126 (tilde ~) only; so this British system can transmit symbols for American but not British currency. How to send that message: 'You owe me #100.' Other TG victims will groan and understand.

TG has another surprise waiting. Businessmen have been known to use the sign @ ('at'), which *is* in the allowable range. If you type it, though, something happens which isn't even hinted in the infamous 'TG Quick Guide'. The entire typed line containing the @ sign is cancelled and not transmitted. What a novelty.

TG charges for basic information about TG. Want to work out how they calculate their incomprehensible bills? First you must register yourself as worthy to learn this occult data (which took us weeks). Then you enter a special command, 'INFO SBI'. Whereupon screen after screen of weird text scrolls by, full of naff layout and # signs, none of it seeming to bear any relation to your bill ... and you pay for every character.

TG lives in the past. When you ask for 'INFO' on something, you are instructed to set the printer paper to the top of the page. They assume you're using not a computer but a teletype.

TG's message services are all mutually inconsistent. There is one almost straightforward way to send a plain message to another user. A substantially different and more tortuous route will send a telex. And a third, insanely complex path must be followed if you want to send a fax, with numerous daft requirements ... e.g. you must enter the *full* international dialling code to fax someone in far-off Britain.

Finally, TG has a sense of timing which all alone is a classic example of bad programming. Sometimes, new messages arrive in your 'mailbox' after you've looked through it, but before you actually 'log off' and leave TG. When this happens, TG issues its usual mass of pointless sign-off information ('CPU time 01 seconds': gosh, thanks), then announces that a message has arrived – 'Mail call (1 unread)' – and *then*, without pausing one instant, it disconnects you and leaves you to plod again through the whole laborious business of getting back into TG, should you want to know what your message actually contains. Grown men have pulled their own heads off in rage at this trick.

(Even Prestel, the cheapo system mentioned in my October 1986 , October 1987 and January 1988 columns – see pages 12, 34, 40 – tells you as you leave whether new electronic messages have just arrived, and gives you the option of staying connected in order to read them.)

Why do I suspect that British Telecom would prefer people to switch to fax? In confidence, what *we're* doing now is posting each other text on disks: nearly three-quarters of a million characters for the cost of a stamp. You know it makes sense.

8000 Plus 56, May 1991

• Isn't it nice that we have this Internet thing nowadays? When the broadband connection is running a little slowly or a favourite website is temporarily down, I console myself with reminiscent thoughts of Telecom Gold.

The Knowledge Trap

'Write about what you know' is the advice given to aspiring authors in a million How To Do It handbooks. Like almost every terse, unambiguous recommendation it needs qualification, and there are exceptions.

Writing about writing is the most seductive trap of all. You are a writer. Day after day you sit there on your bottom, growing ever more pallid and flabby, bathed in the eldritch green light of the PCW screen. Inspiration is needed. A little voice says, 'Write about what you know.' You start tapping out a story about a writer sitting in front of a word processor strangely like yours.

Gosh, this is easy! Drawing on memories of times when you're *sure* that what appeared on the screen wasn't what you typed, you develop a plot about a word processor that *comes alive* and rewrites the author's words, takes over what passes for his or her mind, hacks into the US defence network, starts a nuclear war, becomes God or Satan or even the editor of *8000 Plus*, etc.

Or instead you relax by playing some computer game, suggesting a story about a game that becomes real, or traps the player inside the program, or ...

The trouble is that, leaving out the later and more apocalyptic developments, all these storylines are based on what you know *too* well: sitting at a computer keyboard. Astonishingly, hordes of other writers have already found themselves in the same position and rung endless changes on the above plots. Editors moan and gnash their teeth at the sight of yet another minimally different variation.

(There are non-fiction equivalents, such as the incredibly droll article about hilarious alternative spellings suggested by LocoSpell.)

This aspect of writing about what you know actually predates word processors. The annals of SF are full of strange typewriters: in 1955, for example, Damon Knight published a story about a writer whose typing errors proved to be coded messages from world-dominating supercomputers. The infamous L. Ron Hubbard anticipated the 'trapped inside a computer game that has become horribly real' in 1940, with a story about being trapped in the plot of a hack novel.

Which reminds me of another pitfall pointed out by the critic Nick Lowe in a cruelly funny article. Writers often drink a lot of coffee while tapping out their masterworks. Coffee is much on their minds. Fat, lazily written books thus tend to contain all too many scenes in which, without advancing the plot in any way, the characters stop for a nice cup of coffee. When you start to notice and count these instances of authors 'writing about what they know', it can become downright embarrassing.

There again, suppose you know something that's not commonplace but fairly esoteric. Can it be exploited in a story? Well, if it's such information as how to create highly specialist PCW programs or how to detect the sex of frogspawn, it might be wiser to write it up (if at all) as a straight article for a computer magazine or *New Scientist*, respectively.

The thing to beware of in fiction is the Great Expository Dollop required before lay readers can understand the revelation you plan to spring on them. In bad SF this is traditionally introduced by having someone say, 'Gee, Professor, I know you told me already but how does the frogspawn-sexing widget work exactly?' The Professor then begins, 'Well, son, it's like this,' and talks without interruption for three pages.

Alternatively, the Professor is dispensed with altogether and you get a multi-page chunk of imaginary history starting 'Twenty years ago, the ships of the Federation had ...' and ending with the pungent words, '... and so the free universe was saved and frogspawn-sexing became an exact science.'

No, this story will only move properly if the weird science is woven into its fabric, revealed bit by bit in calculated, teasing asides. And it should be important to someone in the tale. A major character's happiness, or job, or life, must depend on getting that frogspawn properly sexed.

Another dangerous kind of specialist knowledge is the Thing That Really Happened To You. If you're writing straight journalism, the facts are supposed to come first (I know they don't always). In fiction, though, that's not the point.

In writing fiction, you are constructing a narrative machine designed to give the reader particular sensations of excitement, wonder, terror or whatever. If one of the components doesn't fit, it's no good wailing – as inexperienced writers so often do – 'But that's the way it *happened*, I was *there*.' Real-life incidents generally need to be distanced, fictionalized and filed down to shape before they can work as a cog in the narrative. Authenticity, alas, is no guarantee of artistic value.

My last category of dubious knowledge is a more debatable one ... but I think it can be dangerous to be too exclusively well-read in the kind of genre fiction you plan to write.

Take science fiction (as I usually do). Obviously no one has much real specialist knowledge of such familiar SF gadgets as matter transmitters, faster-than-light spacecraft and time machines. But in a nebulous way there is a kind of SF condenses about these and many other strange devices. If you've read all the SF ever written, not only will your brains have turned to soup but you'll find yourself heavily influenced by what past writers have done.

Sometimes, agreed, this can help you avoid hackneyed plot devices. But SF which builds only on other SF will almost automatically emerge as stale, with small twists on established ideas rather than anything brand-new. If you are a towering genius, maybe you can do it. But far easier than being a towering genius is to read a lot *outside* SF, both fiction and non-fiction, in the hope that some strange hybrid notions will emerge from the cross-fertilization.

Ultimately, as you sit there staring into the terrible void of that blank computer screen, you have to coax something new out of the unknown territories of your own imagination. It's then that the advice in the handbooks needs to be rephrased: 'Write about what you don't know.' Or don't know yet ... until you've written it.

8000 Plus 57, June 1991

Desperate Times

Whether or not there's a genuine recession in the world of publishing, the industry seems to have talked itself into one. From my safe bunker I keep hearing the screams of editors toppling from fifth-storey windows with one last rejection slip still clutched pathetically in their hands.

I promise that I tried *very hard indeed* not to giggle when this happened to the chap who last year returned a submission with mild apologies for having sat on it for five agonized, suspenseful years.

Other editors are pale and sweaty after being summoned to the offices of Higher Management, shown the vertiginous drop from the window-ledge, and told, 'One day, son, all this could be yours.' Then they are released with a caution and instructions to spend less. Except for a guaranteed international best-seller, in which case they can offer as many millions as they see fit. It's difficult just now to be an editor without spending hours gibbering and biting one's toenails.

These symptoms are caused by such decisions as: do you lash out £2,000,000 trying to poach a horror megastar author who always does well? It's relatively easy for editors to sell this idea to the holders of the purse-strings: they can point to a proven track record of best-sellers. Unfortunately eight other publishing houses are planning exactly the same. By the time the shark-like bidding frenzy is over, the deal is made, and the book is launched with the major publicity campaign demanded by such an investment, it's possible that the result will be that common phenomenon of the 1980s – the best-seller *which loses money*.

This happens because today's surreal figures in the genre of Fat Books With Embossed Foil Jackets are such that, quite often, there is no reasonable possibility of making enough profit from the book to cover the initial advance to that megastar author. The publishers are in theory aiming for a long-term investment: goodwill or contractual ties which will keep the author with them, and hopes that this publicity splurge will cast a glow of increased desirability over his or her later books (plus earlier ones which they will try to buy in economically and reprint).

Of course, the moment the megastar is off that contractual hook, the frantic auction could start again, with a different publisher winning the right to pay X million pounds ... and bye-bye to the first one's 'far-sighted' investment.

The notion of long-term goodwill used to apply to first-time authors. Although there are exceptions, a first book tends not to do very well. Editors used to regard it as an investment, with hopes of winning the loyalty of promising authors. A modest advance would be paid, a modest print-run scheduled: three or four books hence, this could be an established writer. And the investment was a small sum which unlike those speculative seven-figure advances couldn't possibly threaten the whole publishing house.

This is the paradox. Some editors try to woo promising authors exactly as I've described, but find it harder and harder to 'sell' their decisions to an editorial board which keeps screaming, 'We've *got* to have another best-seller!'

It's notorious that committee debate issues in inverse proportion to the sums involved. A multi-million-pound expense tends to be nodded through, while an order for paperclips will split the meeting in bitter controversy about expense, waste, possible alternatives, and the villainy of staff who don't appreciate that these luxuries cost £1.20 per thousand.

Just so, our editor finds it incredibly hard to make the case that Fred Obscure is worth a modest investment. There is a hiss of indrawn breath around the boardroom table. Think how many *paperclips* you could buy for £3000. 'What's his track record?' snaps a thin-lipped editorial director. First-time novelists do not tend to have much of a track record.

(Exceptions are made for media personalities, noted prostitutes, famous terrorists, and so on. They may not be able to write, but get their X on the contract and the copy-editors will sort it out. An editor once boasted to me that he'd ghosted a major bestseller in four days, based on fragments of the 'unpublishable' typescript from a famous name I'm not at liberty to divulge.)

As if the above picture weren't depressing enough, there's a growing possibility that submissions from new authors might never get read at all. With the latest waves of staff cuts, editors have less and less time to scan everything that comes in.

Traditionally, the first sifting of the slushpile is performed by outside help – by freelance publishers' readers, a particularly overworked and ill-paid breed (described here in December 1987). They're now having a thin time of it, many being effectively laid off. Some major paperback imprints are no longer employing readers at all ('what, pay *freelances* in a recession?'), while the depleted ranks of editors still lack time to do the work. Never before in the history of literature have so many works been rejected unread by so few.

(Personal note: your columnist was recently privileged to read the entire popular-science slushpile of a noted publisher, a terrifying stack of cobwebbed manuscripts that had been accumulating for six months. 'You have one afternoon to report on the lot,' they said. Half a day's pay was all the budget would allow. I staggered home with Gumby-like groans of 'Urrr, my brain hurts ...')

So what hope remains for all the aspiring novelists tapping away at PCWs in a corner of the bedroom? There are two encouraging points to bear in mind.

One is that there's something artificial about the current panic. World upheavals may send dread ripples down from the vast conglomerates who directly or indirectly own many British publishers ... but the general public hasn't stopped buying books, you know. In six or twelve months' time there could be an equally wild swing to optimism and a boom might even be declared. In which case some publishers will look a bit silly, like shipowners who jettisoned half their valuable cargo at a premature storm warning. If so, they'll be keen to build up their lists again. This means you.

The other point is that most really good stuff *is* sooner or later pushed through the horrors of the system by an editor who believes in it. Today the chances are fewer and the work needs to be better, but that loophole is still there.

We live in interesting times.

8000 Plus 58, July 1991

Going Commercial

After months of perfectionist polishing, your program is brilliant. Anybody with a PCW and more than three brain cells has been yearning without knowing it for this wonderful software. If marketed, the product would practically sell itself. How to start?

Having entered the shallows of commercial software in a struggling two-man partnership, growing over the years to a struggling limited company with fewer than three employees, I know roughly how it works. None of what follows is intended to be depressing.

First ... the world must hear about the amazing product we'll call SuperProg. (Silly capital letters in unlikely locations are the essence of software street credibility, says pundit LangForD.) This means advertising in magazines, and not just once: too many floating readers buy only the odd copy and have an eerie, psychic ability to pick the month you tried to save money. Your eyes will water when you hear what ad space costs, but this may be negotiable.

Some magazines will typeset your ad to order, which could be useful if you lack a desktop publishing kit and are not a whiz with Letraset. On the other hand, in 1991 everyone knows someone with access to high-quality DTP – don't you? If the quality isn't too high, a good ploy is to produce copy at double or triple size: photoreduction will help fine up those ragged lines.

It is traditional to send out review copies of software. It is also traditional for magazines to be short of space just then, but you never know ... worth a try.

Before this launch, and in readiness for the immense rush of orders, you need to produce a SuperProg manual. Even if the software is self-explanatory, most people like the reassurance of clear, literate and professional-looking instructions. Of course you'll have no trouble with the first two criteria. Tip: borrow someone who knows nothing about your software, and see how he or she makes out with the draft manual.

'It says I should exit from the widget menu. How?' 'Press Exit, of course.' 'You don't say so.' 'But isn't that *obvious*?' 'No.'

Another tip: while the manual need not be boring, no one will praise you for filling it with naff jokes. Imagine if you had to suffer the Japanese sense of humour every time you puzzled over the VCR manual. (Some people suspect that this is in fact the case.)

Modern technology smiles on small print runs. A High Street print shop can whiz off just 20, 50 or 100 photocopies of the book, quite possibly with automatic collation thrown in – so you needn't sort the sheets into booklets yourself. In the not-so-distant days of photolithography, it was tempting to order too many copies since the unit cost dropped so rapidly after the savage expense of setting up.

Nowadays offset litho starts getting competitive at around 400-500 copies, or more. Always ask first.

Or perhaps you have access to a photocopier. Beware of overstraining small desktop models, or of relying on any but hefty commercial copiers (sometimes not even them) for large-volume printing – especially on both sides of the paper.

How to bind the books? 'Traditional' American 3-hole ring binders are awkward-sized and expensive both to buy and to post. Over the years we've used A4 folded and centred-stitched into A5 booklets; then two-hole-punched paper in flat plastic binders; and finally a 'perfect binding' system. Here you drop the pages into prefabricated jackets with a strip of glue along the spine: a heater melts the glue and when it sets it's a book. (Plug: Heyden 'Bind-It', Spectrum House, Hillview Road, London, NW4 2JQ.) Spiral binding is now cheap, I hear.

Disks are bothersome. The PCW 3" size is more expensive and less discounted than any other; you can get some reduction by buying hundreds, but might as well go for the best mail-order price you can find.

Should you register for VAT? *Pro:* you pay VAT on nearly all your supplies (the manuals might be an exception if the printers accept that they are VAT-free 'finished books' or 'newsletters') and would be able to reclaim this – also on the relevant percentage of your phone bill, of which more below. *Con:* you will have to charge VAT. Thus if you reckon that £20 is the least gross return you can bear to rake in from each package, you must add 17.5% and advertise it at £23.50, doing the books and paying all the odd £3.50s to Customs & Excise each quarter.

Should you form a partnership to market SuperProg? *Pro:* you share the work and responsibility. *Con:* tax inspectors are legendarily suspicious of partnerships, and your dealings will need to be sanctified by paying an accountant. (Well, that's what *our* accountant told us. Virtually all accountancy firms are partnerships, so they should know.)

How about becoming a limited company? *Pro:* small companies are often formed as a kind of insurance. Suppose your product SuperProg turns out to be the trademark of some US outfit, and they sue for the traditional million dollars. No matter how you grovel your way out of this, there's a chance of an unfunny legal bill which might cost you your house (say). In this remote emergency, liability would be confined to company assets and you at least wouldn't go down the hole.

Con: in exchange for this legal protection, you must pay hefty annual company fees, submit annual reports to Companies House, pay your accountant more as official auditor, and run your business along particular lines (no more dipping into the profits whenever there are some ... you need to put yourself and co-directors on a salary, and deduct tax via the PAYE system).

In the least-known Gilbert and Sullivan operetta, *Utopia Ltd*, everyone becomes a limited company in order to get out of paying bills. In the real world, this is what the arcane jargon of lawyers and accountants would term a definite no-no.

Now, a big one. Think carefully: will you publish your phone number? If not, and especially if SuperProg has a genuine or apparent 'rival product' with a contact number, you lose sales dramatically.

If you publish it ... are you in all day every day? Better invest in an answering machine. Want to use your phone for social purposes in the evening? A separate line *and* an answering machine. Do you think the phone number is for enquiries only and that software problems can best be handled by mail? The people out there think differently. Maybe a separate line and a fax machine, to keep things impersonal ... but not everyone uses fax (this might be different before long) and you lose sales. How about that Amstrad fax that turns into an answering machine for voice calls? Not cheap ...

And to give technical support, you must learn tact. An expert gave me this example:

Wrong: 'Is it plugged in?' The instant answer: 'Of *course*! Don't be silly.'

Right: 'Some computers went out with defective mains plugs. Just to check, can you read me the BS number printed on the inner face of the plug?' The muffled answer: 'Hang on ...' (bump, clatter, pause) '*Oh*, I think I see what's wrong ...'

Welcome to the wonderful world of business.

8000 Plus 59, August 1991

The Padded Cell

From time to time I'm expected to do the decent thing and reveal the ultimate secret of literary success. Unfortunately there are hindrances to such altruism.

The first is that if I had the Secret, I wouldn't be disclosing it here but would be lolling in luxury atop the best-seller list. (Which in the present recession might not mean much. Someone recently revealed that one of the 'top 100 bestsellers' in a particular chart had sold precisely one copy.) It's like those programs that tell you how to grow rich by betting on horses ... you wonder why a programmer with this Midas touch is bothering to sell software for trifling sums, and why thousands of bookies aren't going bankrupt. Perhaps *they* buy these packages too, and use them to adjust the odds.

The second snag is that there is no one literary Secret. Instead there are hordes of small insights, none particularly valuable alone. Think of chess, whose rules are far simpler than those of English grammar and which (mathematically) allows ridiculously few possible games when compared with the number of legitimate ways of arranging words into a book. Yet even there the Big Secret is unknown, apart from platitudes like 'Play better than the other guy.' There are rules of thumb: aim for control of the centre, avoid doubled pawns, etc. But ultimately you have to win by playing well, which seems undefinable. Like writing well.

This month we study another little facet of writing – in fact an ancient and powerful literary technique which you never see mentioned in creative writing manuals. Here's a textbook example from Britain's most prolific science fiction writer, Robert Lionel Fanthorpe:

> ... The silence was broken by metallic noises. Harsh clanking, jarring metallic noises. Things were stirring within the disc ship. Strange metallic things; things that were alien to the soft green grass of earth.
>
> Terrifying things, steel things; metal things; things with cylindrical bodies and multitudinous jointed limbs. Things without flesh and blood. Things that were made of metal and plastic and transistors and valves and relays, and wires. Metal things. Metal things that could think. *Thinking metal things*. Terrifying in their strangeness, in their peculiar metal efficiency. Things the like of which had never been seen on the earth before. Things that were sliding back panels.... Robots! Robots were marching ... Robots were marching ... (from *March of the Robots*, 1961)

Yes, the literary technique we're talking about is padding. Appalling passages like the above are favourites for readings at SF conventions, attracting the same groan-laden interest as the immortal verse of William McGonagall. Hypnotized by dreadfully predictable rhythms, audiences have been known to chant along: 'The city slept. Men slept. Women slept. Children slept. Dogs and cats slept.' (*Ibid.*)

In justice to Fanthorpe, I should point out that he was and is an engaging, intelligent fellow. What forced him to these extremes was the job of providing almost the entire SF/fantasy output of the ill-famed Badger Books, in his spare time: 121 volumes over eight years, peaking at the rate of one per weekend. Gasp.

A less than funny point underlies all this. The typical paperback best-seller is a big fat book containing all too much padding. Please note the word 'typical': I'm not trying to state a universal law here. There are always special cases. I'd hate to have to edit a word from the fat but densely textured *The Name of the Rose*.

Many such evils are committed in the name of research. It's easy to stuff a historical blockbuster with wodges of information torn raw and bleeding from history books. Nautical adventures bulge with stuff about splicing the mizzen-fo'c'sle's tween-yard quoins and hurling loblolly on the futtock ratlines (or words to that effect), lifted boldly from textbooks or C.S. Forester. War/thriller epics revel in the clotted jargon of boys' toys:

> Lovingly he hefted his favourite automatic gun, an idiosyncratic VK 155 L/50 Swedish equalizer with 155mm calibre and a taut 865m/sec muzzle velocity. His mates scoffed at its bulk, but the chunky 14-round magazine and 25km range made it his kind of weapon, tough and mean....

There you are, all done from two minutes with *Brassey's Artillery of the World* – which gives enough information for several more pages, including the interesting fact that it's a self-propelled job weighing 51 metric tons. Oops.

This leads to the much-feared Brand Names ploy. Ian Fleming made it famous in the Bond thrillers, which are too tightly written to be accused of padding. The idea is that judicious mention of the brand names surrounding your characters will economically anchor the scene in reality *and* tell you something about the people. Snob appeal is the easiest route: Gucci luggage, Yves St Laurent socks, or whatever. SF writer William Gibson played a neat literary con-trick by extracting the same knowing effect from *nonexistent* brand names, writing about the Ono-Sendai cyberspace deck in a way which made it clear this is the model the real in-crowd uses....

Now it's all escalated. You get long lists of names, descriptions of shopping for *more* brand names, descriptions of cars, cameras, wallpaper, bedding, clothes and much more. Amid these branded props, characters go through old routines from when writers were paid by the word: fiddling at length with cigarettes and cups of coffee, discussing the plot again and again in repetitive dialogue ... fill in your own list.

Why? I have a nasty suspicion. The computer world is full of people who buy games which present an interesting challenge, and then buy information on how to cheat so there *is* no challenge. 'Cheating' in books is easier: just skim. G.K. Chesterton made a distinction between 'the eager man who wants to read a book, and the tired man who wants a book to read'. Tired readers demand a fat book to pass the time, but skim inattentively. Padding and repetition come into their own, almost as though designed for sloppy reading. What a horrid theory. Can it account for Jeffrey Archer?

There is in fact a constructive use for padding. Unless you have the stamina of a Lionel Fanthorpe, producing sheer waffle is actually very exhausting. A far better SF author, Robert Sheckley, claimed this as his infallible cure for the terror of writer's block, when nothing will come. (Asked if *he* ever suffered this, the hugely prolific Robert Silverberg said: 'Once. It was the worst ten minutes of my life.') Sheckley tried a drastic exercise to get things flowing again: making himself type 5000 words a day, any words so long as he met the quota, grimly bashing out stuff like ...

> Oh words, where are you now that I need you? Come quickly to my fingertips and release me from this horror, horror, horror ... O God, I am

losing my mind, mind, mind ... But wait, is it possible, yes, here it is, the end of the page coming up, O welcome kindly end of page ...

After days of this, Mr Sheckley made the great discovery that it was now actually easier to write a story than go on suffering. So he did – saved by padding, though at such cost in mental anguish that he vowed never to use this ultimate literary secret again. It's all yours.

8000 Plus 60, September 1991

Losing Your Chains

One by-product of the 1991 silly season was a spate of circulars, many recognizably LocoScripted on the ubiquitous PCW. I suppose there's no way to prevent nutters and con artists from buying word processors, just like the supremely law-abiding readers of this magazine....

The most irritating item was a long tract offering a very expensive book on how to cheat. Its amazing technique of 'neocheating' would enable you to win at cards, rip off friends and business associates, and rule the world – all in a completely safe and undetectable way. Of course, the brochure adds, you don't *have* to cheat at cards. The idea is to attain inner tranquillity and control, stuff like that. Being able to steal your neighbours blind will be a mere side-effect of your astonishing new powers.

(Remember those old martial-arts adverts? 'After ONLY THREE LESSONS you will be able to WRENCH OUT PEOPLE'S INTERNAL ORGANS and REDUCE THEM EFFORTLESSLY to a BLOODY PULP. Your new-found ability to MAIM, HURT and KILL should be used only in self-defence....')

A more familiar arrival was a posh, word-processed version of that old standby, the chain letter. I thought they'd died out years ago. Ah, the schoolday nostalgia of sending a nice picture postcard to the top name of the six listed, plus four plain cards to friends with an exhortation to do likewise and not break the chain. Then you waited for the glorious bounty promised by the logic of mathematics when your name rose to the top: 4 to the power of 6 or 4,096 cards, probably from exotic places all over the world!

I never got any, which was just as well since on reflection I had absolutely no idea what I'd do with them.

Today's chains have progressed beyond the simple innocence of long ago. The example I received is full of claims that this is not a chain letter but that entirely different and legal thing, a Multi Level Mail Order Sales offer. The currency unit is no longer the humble postcard but the used £5 note. (No cheques.)

Now, when you send fivers to the previous names on the traditional list, you are supposedly buying four valuable 'financial and business reports'. Since these consist of at most two sides of paper, you can imagine how expensive they are to produce, and how crammed they must be with vital data.

But it's unnecessary for you to read these ludicrous texts. All you do is pay a fiver apiece for the things, and photocopy them for sale to the further suckers whom you yourself are expected to rope in. 'It is legal because you are offering a legitimate product to your investors.' See? In due course, as the circle of golden opportunity expands exactly like a chain letter ('imagine if everyone sent out 1,000 letter packets!'), you in your turn are promised tons of fivers.

There's something slimy and repellent about the several sheets of 'don't miss this great chance' evangelism which landed on my doormat. Naff testimonials from supposed past operators of the scheme are included: why do they all quote vast profits in pounds but write suspiciously like Americans? Any guesses as to where this con originated?

Let me make some obvious comments. The first concerns a testimonial from the 'originator of this plan', who is letting others into it out of sheer benevolence, having himself retired long since on claimed profits of over £4,000,000. Stop and think. If we're to believe him, every penny of that sum must have come from hopeful 'investors' who paid £20 for four joke reports at a fiver each. Gosh, there must already be *two hundred thousand people* on this game! If each sends 1,000 further packages as recommended, there are two hundred *million* in circulation already. A bit late for anyone else to jump aboard....

All right, most of the build-up is a pack of lies. My second comment is for people who think a bit further and argue: 'Yes, it's a scam. But if this one is just starting, I could get in now and make a quick profit before it falls apart.' Possibly. But the law of conversation of mass/energy applies to money. Whether the chain letter dies out by expanding to swamp the world or by deserved shrinkage as people chuck the thing in the bin, each fiver you make is sooner or later balanced by the 'investment' of somebody who won't get it back ... someone, perhaps, who believed in this nonsense as a genuine business opportunity, and can ill afford even small losses. The mere fact that you're unlikely to meet these victims of the con is no excuse.

A final cheeky thought. When I mentioned conservation laws, I didn't take into account the fact that real-world systems leak money at the seams. Just so here: all that photocopying or Loco printing, all those envelopes and especially (as the letter insists) *first-class* stamps. Elementary, Watson: the originator of this game must be the one outfit which makes money from it no matter what ... the Post Office!

The above is all small stuff. Over in the wonderful USA, everything is bigger and better, and this kind of scheme can be played for high stakes. One that's caused much trouble in recent years is the Aeroplane Game ... or rather, because they're Americans and don't know any better, the Airplane Game.

This is a plain (sorry) and simple pyramid scheme. I don't care to give full details, but typically the suckers join an imaginary flight as a 'passenger' for $1,500, and by recruiting new victims rise through the ranks of 'flight attendant' and 'co-pilot' until they reach 'pilot' and receive $12,000. For this to happen, 64 more people must have signed on at $1,500, each needing to recruit 64 more before *they* collect ... and so on. Since several times the world's population will soon need to join and pay, a rude crash-landing awaits overwhelmingly many of these clowns.

(The Airplane Game is illegal in numerous US states, including New York, Texas and California, which have laws against dodgy pyramid schemes.)

Am I just repeating myself? The creepy aspect of the American swindle was how it raged through the unworldly communities of what is called the New Age movement. That it was crass, immoral and doomed was evident to anyone who chose to think hard for a moment: but the open-minded suckers swarmed aboard, dazzled by 'spiritual' jargon about 'infinity processes' and 'abundance workshops' which promised that free money could be spun out of nothing, forever.

After all, the sceptics who rudely told them the harsh facts of mathematics had been equally disrespectful about vital New Age concepts like astral channelling, reincarnation and magic quartz crystals whose 'energy flow' can cure AIDS. Who would listen to horrid people like that? Who indeed.

Usually this page talks about writing. This time we've sneaked around to it from behind: as George Orwell said again and again, to write clearly you must first think clearly. Myself, I'm every bit as lazy and greedy as the next PCW owner, but I have learned to stop and think about certain things. Writing is one ('did I really mean to say *that*?'). Get-rich-quick schemes are another.

8000 Plus 62, November 1991

Old Joyce's Almanack

1 January: New Year's. All over Britain, professional writers resolve that in 1992 they will be kinder to the often reviled tribes of book publishers and editors. After all, are they not human? If you prick them, do they not bleed? At the same time, computer journalists resolve never again to abuse Alan Sugar: is he not human? If you prick him, does he not sue?

2 January: Writers lose patience and start to draft intemperate letters beginning, 'Where are the royalties that were due in April 1991, you bastards?' Computer journalists grow bored and sketch out headlines like AMSTRAD'S BEARDED LAGER LOUT IN PCW BONK HORROR, hoping that a story to fit will sooner or later turn up. (Recommended reading for those who've read too many headlines like that, and even more for those who write them: *Waterhouse on Newspaper Style* by Keith Waterhouse, which is good on the distinction between striking, memorable and just plain awful tabloidese.)

February: Public Lending Right Month. Authors rejoice to receive generous payments for loans of their books from public libraries. Yes, every single borrowing (except the ones from libraries not actually being monitored by the PLR scheme) brings the lucky author a fraction over one penny! To understand the colossal esteem in which our country's writers are held, you have only to compare this sum with the typical cost of, say, hiring a video....

March: National Hackers' Week. There is a US program called Password Coach, sold as a security measure to stop the users of big corporate computer networks from choosing passwords which are just too predictable by evil-doers. The list of 'cultural icons' from which all too many Americans pick their passwords includes Alka-Seltzer, Aristotle, Asimov, Beatles, Brando, Disneyland, Dracula, Frodo, Garfield, Godzilla, Hitchcock, Hitler, Jesus, Jetsons, Kirk, Kleenex, Madonna, Nintendo, Scud, Superman, Wheaties, Xerox ... (Reported in *Harper's Magazine*, November 19 91.) Inspired by this insight, a group of hackers attempts to break into Amstrad's secure internal network using the passwords JOYCE, PCW, SUGAR and SHIFTEXTRAEXIT, achieving final success when they key in the word MONEY.

(Fact: when I was doing wicked things on an Oxford University computer years ago, a surprising number of passwords proved to be either PEANUTS or SNOOPY – if not PASSWORD. Fact: an anonymous civil servant, working on a nationwide Government computer system that handles billions of pounds, was told of all the above and almost wept at the variety of passwords open to Americans. 'We're only allowed four letters....')

1 April: The Feast of All Missing Address Marks, a day of immemorial rustic pursuits. The colourfully clad address-mark hunters track their fleeting quarry across huge expanses of wasted time and corrupted data blocks, with hearty, traditional cries of 'View halloo!', 'Tally-ho' and 'Forgot to make a backup!'. If a 'kill' is made, huge stirrup-cups are drunk with heartfelt toasts to Dave's Disk Doctor Service. If not, even huger cups are drunk to blot out the thought of retyping the whole document from memory in the days to come.

1 May: May Day. Fresh from a campaign against the pagan and/or erotic associations of traditional May celebrations (especially maypoles), the Fundamentalists Against Absolutely Everything crusade turns its attention to the dubious symbolism of thrusting disks into PCW drive slots. A new PCW add-on reaches the market: frilly covers for disk drives, to help protect the sensibilities of the pure in heart.

June: Still stricken with poverty, the scattered ex-Soviet republics place giant orders for the most economical word processing system available from the running-dog capitalist West. Hard-currency payments for the 'PCWski' being something of a problem, Amstrad reaches a cunning compromise and is soon in the thick of a massive publicity campaign for its new line – forty-two million gallons of own-branded vodka, now available at every branch of Dixons as 'disk head cleaning fluid, nudge nudge, wink wink, know what I mean?'

4 July: American Dependence Day, dedicated to all the owners of incredibly expensive IBM and Macintosh computers whose costly, imported software includes alien spelling-checkers, deluding the hapless buyers into writing 'airplane', 'color', 'honor', 'humor', 'labor' and all the rest. Special naff points are scored by those dim bulbs who try to correct this without actually bothering to use a dictionary, producing nonce-words like 'humourous' and 'labourious'. Recommended reading: *Mother Tongue: the English Language* by Bill Bryson – who, as an American who's lived over here since 1977, has a good transatlantic perspective. *[But still gets things wrong.]*

August: the Silly Season. 1992 sees a spate of new graphic design programs to help PCW owners devise and implement their very own crop circles. Students of the paranormal are soon thrown into uproar by an amazing series of Wiltshire cornfield patterns in the exact shape of a PcW 9256. 'It is completely impossible that this could be a hoax,' gasps a top expert: 'All this week's phenomenal events were accurately predicted by Nostradamus back in the 16th century, in his famous prophetic verse that begins: *A great king of terror (i.e. Alan Sugar) will descend (i.e. falling Amstrad profits) from the skies (i.e. satellite TV), In the year 1999, seventh month (i.e. August 1992) ...*'

September: Sprouts Day. Hot from Brussels comes the long-awaited EC directive on disk sizes, an awesome tome running to 750,000 words. The gist is that to eliminate incompatibility from future computer systems, all 3" and 3½" floppy disks will now be superseded by the new standard Eurodisk, measuring an easily remembered 86.5 x 94 millimetres. This avoids any possible charges of company or national favouritism, by being unusable in all existing disk drives.

After buying up a huge stock of the now-obsolete 5¼" drives for a mere song, Alan Sugar amazes the world by releasing his spanking-new pCw 9257 ... and everyone gets rich selling file transfer systems for moving documents from 3", 3½" and Eurodisk formats to the new Amstrad 5¼" standard.

31 October: Hallowe'en. Strange eldritch Things flit through murky air and stalk the haunted night, being the usual rumours of LocoScript 3 with artificial intelligence, LocoScript 4 with the Virtual Reality interface, etc.

5 November: Bonfire Night, when huge pyres of never-opened Amstrad manuals are constructed in every town and the person responsible for the stupidest computer problem is burned in effigy by technical support departments all over the country. This year's winner is an inexperienced secretary who complains of steadily declining screen visibility when using LocoScript 2. A crack support team from Locomotive finally cures this difficulty after four arduous hours scraping off the Tapp-Ex.

31 December: New Year's Eve. Where does the time go? Your columnist resolves to be nicer in 1993 to editors, publishers, taxmen, VAT officials *and* Alan Sugar....

8000 Plus 64, January 1992

Ticking the Boxes

Once upon a time I wrote the book review column for a fantasy games magazine. Although the magazine used little fiction, and although a lowly columnist should never be confused with that godlike being the editor, readers started submitting stories to me. Being tremendously humane, I devised a form-letter of reply which required inspection of page 1 only. It went more or less as follows. (Since then I've stopped being tremendously humane – no MSS, please!)

Presentation

A tick in this section means that a jaded editor would probably reject you without reading any actual words.

☐ It's hard reading. You need ☐ a new ribbon, ☐ a new print-head, ☐ letter rather than 'draft'-quality print, ☐ larger print (12 letters per inch minimum).

☐ Only black ink is acceptable. Keep that violet ribbon for personal friends, or enemies.

☐ Line spacing must be set to 2.

☐ Margins too narrow: leave at least an inch all round.

☐ The MS lacks ☐ your name, ☐ your address, ☐ a word count ('5,341 words' brands you an amateur – round it to the nearest 10 or 50), ☐ a traceable title, ☐ page numbers.

☐ Paragraphs aren't (or aren't clearly) indented. Set the tab to at least half an inch.

☐ The paper gives a poor impression: ☐ too flimsy, ☐ grubby from previous submissions, ☐ unseparated continuous stationery.

☐ Too many obvious hand-corrections – three per page at most. (Don't be a cheapskate ... reprint the page.)

☐ Flawed punctuation: ☐ spaces are needed after commas etc., and ☐ are not acceptable immediately before.

☐ Don't break words over line ends by inserting hyphens – printers hate this.

☐ Use paperclips or staples to hold the pages together (in separate chapters for long works): elaborate bindings are frowned on.

Minor points: ☐ editors prefer bare-minimum covering letters and do not wish your opinions on the brilliance of the story; ☐ right-justified text is generally disliked (but more editors are coming round to the use of actual italics, in place of indicating them with underlines); ☐ page numbers are best placed at top right, in a header including your name and perhaps an abbreviated title; ☐ leave the top third of page 1 blank for title layout design.

Skimming the Opening

No single item here should drive you to despair, though a bad mark for spelling should come close....

☐ The title is ☐ boring, ☐ clichéd, ☐ pretentious, ☐ incomprehensible, ☐ an over-familiar quotation, ☐ already used by a better-known author.

☐ The opening is ☐ uninspiring (doesn't encourage anyone to read on), ☐ clichéd, ☐ pretentious, ☐ incomprehensible, ☐ over-melodramatic.

☐ The spelling is dodgy. Writers should refer constantly to the dictionary. If you use a spell-check program, beware of ☐ adding your own wrong spellings to its lexicon, or ☐ accepting typos merely because they're passed as words – are they the *right* words?

☐ Dodgy grammar and syntax. Read more books.

☐ Dodgy use of punctuation. Worst common faults: ☐ contracting 'it is' to 'its', ☐ using 'it's' as a possessive, ☐ putting spurious apostrophes in plurals (e.g. 'plural's'), ☐ using commas where another link like a semicolon or colon is required. (Or a full stop.)

☐ Proper names and fresh sentences should start with capitals. Caution and restraint are advised when you're tempted to use caps for Ironic Effect or ATTEMPTED EMPHASIS.

☐ Don't keep italicizing your dramatic or funny lines. (This over-emphasis gives the same effect as laughing loudly at your own jokes.)

☐ The opening features needless explanations, awkwardly inserted lumps of information, or fussy footnotes. Be subtler.

☐ The sentences are generally too long, defused by add-on clauses and afterthoughts.

☐ Too many non-sentences, odd qualifying clauses that don't join on to anything.

☐ Over-use of stock SF jargon. (This and the following reflect the submissions I used to get!)

☐ Over-use of ye olde fantasy fustian style. Long strings of sentences starting 'And' or 'But' are a dead giveaway; likewise a plethora of those all-purpose Tolkien adjectives like *chill, cold, dark, dread, lone, pale, wild,* etc.

☐ Over-elaborate style: the reader can't see the story for the words.

☐ Over-bald style: a distinct lack of atmosphere or sense of place.

☐ Too many atmospheric adjectives and adverbs: indirect evocation via metaphor, simile (in moderation) or mere choice of words can be far more effective.

☐ Needless avoidance of the word 'said'. Trying for freshness by writing *'Hello,' he yawned; 'Goodbye,' she exploded* gets mechanical and silly. ☐ In a long *he said/she said* exchange, several 'he/she said' phrases can be omitted once the rhythm of dialogue is established.

- ☐ 'Make dialogue sound like talk, not writing,' said Wolcott Gibbs of *The New Yorker*.
- ☐ Character names too ☐ joky, ☐ routine (all Smiths and Joneses), ☐; clumsy, ☐; somehow familiar.
- ☐ Initial situation seems far too familiar. (Surprises overleaf? First you have to make the reader turn over.)
- ☐ Awkwardness with point of view. Who's telling the story? If the narrative voice is one character's, beware of omniscient commentary from the author ... and vice versa.

A Footnote: Practically every 'fault' above has been committed by a great writer. Genius can excuse many things, and can even make a virtue of what for most people would be flaws. But are you *sure* that you're a literary genius?

In Summary

This is where I'd finally confess my snap judgement:

- ☐ No serious problems were found in this necessarily brief sampling. Try it on a real editor. Or ...
- ☐ A professional editor probably wouldn't have finished page 1. Keep trying, but not on me.
- ☐ You enclosed a suitably stamped and self-addressed envelope, and the MS is returned herewith. (Was the postage insufficient? Your problem, mate.)
- ☐ No postage enclosed. MS chucked in bin. This has been a stern lesson from the school of Life.

A Final Note: Some of the judgements that I'd make when ticking items on this form were of course a bit subjective. But if you're an aspiring writer, try going through some piece of your own fiction while pretending to be an utterly merciless, unsympathetic swine (me), and see how many ticked items you can amass by never giving yourself the benefit of the doubt. The results might be instructive.

PCW Plus 66, March 1992

False Prophets

In between *PCW Plus* columns and software work, I am a science fiction writer. It is an awesome responsibility. SF writers have a duty to prophesy the future: everyone knows that. Keen-eyed peerers into the awesome vistas of futurity, that's us.

Well....

One of my all-time favourite openings to a book is G.K. Chesterton's 'Introductory Remarks on the Art of Prophecy' in *The Napoleon of Notting Hill* (1904), which in one of those odd literary coincidences is set in a future Britain of 1984. This begins: 'The human race, to which so many of my readers belong ...' and continues by outlining the rules of humanity's favourite game, known as Keep Tomorrow Dark, or Cheat the Prophet.

'The players listen very carefully and respectfully to all that the clever men have to say about what is to happen in the next generation. The players then wait until all the clever men are dead, and bury them nicely. They then go and do something else. That is all. For a race of simple tastes, however, it is great fun.'

Chesterton goes on to explain that the game became very difficult in the twentieth century, because there were so many clever men making prophecies (especially H.G. Wells) that it became very, very difficult to do something that hadn't been predicted....

This is of course the secret of success in SF prediction. Make enough wild guesses and one or two of them must surely hit the mark. Bingo, you're a prophet.

The next logical stage is for SF authors to develop delusions of grandeur about this, and to start judging their betters by the daft yardstick of successful prophecy. Take a bow, Isaac Asimov – who in his book *Asimov on Science Fiction* spends several pages putting the boot into George Orwell for getting it all wrong about what life was going to be like in 1984. Orwell failed to predict computers, to foresee the oil crisis, to invent exciting new science-fictional drugs and vices, etc. Silly man! Anyone would think that he was neglecting his duty as prophet and instead writing a savage metaphorical fable about 1948. (Orwell himself seems to have been under this foolish impression.)

It's strange how many people remain convinced that SF writers are all at least *trying* to foretell the One True Future, if not to bring it about. The world a writer might most like to live in would probably be too luxurious, hedonistic and devoid of tax inspectors to allow exciting conflicts or plot tangles ... so most SF has settings which are interestingly flawed, if not out-and-out awful warnings. To pick one example from many, Sir Kingsley Amis was thoroughly clobbered when he told the dread Margaret Thatcher that his novel *Russian Hide and Seek* (1980) was about a future Britain crushed by Russian occupation. She blasted him with: 'Can't you do any better than that? Get yourself another crystal ball!'

In fact, as you might imagine, SF has a fairly poor record of prediction in the high-tech world of computers.

One SF trend was that of the giant brain. Gigantic mainframe computers were a fairly safe bet by the time of the Second World War. and so they proliferated in stories. A.E. van Vogt's computers were called the Brain or the Machine. Isaac Asimov's was called Multivac (a bit of a slip there – early computers tended to have names like MANIAC for Mathematical Analyser, Numerical Integrator And Computer, but Asimov made the AC in Multivac stand for 'analogue computer' – rather than the *digital* computers which came to rule). Lloyd Biggle Jr opted for size with Supreme, which ran the galaxy and filled an entire planet. Other writers stuck to elaborate mechanical cams, or wrote about astrogators fiddling with slide-rules while the whirring and clicking of electrical calculators filled the space-ship's bridge....

All this clunkiness was understandable in the light of what has passed for a personal computer within my own lifetime. The first home computer manual, so far as I can trace, was the 1955 *GENIACs: Simple Electronic Brain Machines and How to Make Them* by Oliver Garfield. In these crude logic circuits, 'hard-wiring' was the rule. The 33 ultra-simple 'programs' that your GENIAC could run actually demanded that you rewire the circuits for each program change ... in effect, building 33 slightly different machines. (My thanks to Nigel W. Rowe for this information.) Those were the days!

Meanwhile, SF's other big thing in the computer line was the walking, talking robot, still conspicuously not a feature of everyday life. A robot that can wander around getting up to mischief clearly has far more dramatic potential than a dirty great lump of circuitry. The writers weren't interested in prophecy: they wanted to tell thrilling – not to mention startling, amazing and astounding – stories. And so the robots began to march, spreading havoc and cliché in their pitiless wake.

But then, there are such things as self-fulfilling prophecies. We are all somehow fascinated by the idea of thinking robots that walk like human beings. Perhaps one day the stories will be said to have come true. Even now, a lot of the enthusiasm for Virtual Reality systems seems to have been fuelled by the 'cyberspace' SF visions of William Gibson and his merry band of imitators. Life imitates Art.

Meanwhile, with thousands of SF writers letting their imaginations run riot in disgraceful orgies of prediction, a few will always diverge from the trend – which is lucky for them if it's a dubious trend, like the too-easy glossing over of the incredible difficulties in creating intelligent robots.

A possible minor success in the prophecy game was scored by Anthony Boucher, whose 1942 story 'QUR' (Quinby's Usuform Robots) started with the received wisdom of robots as imitation people, and suggested that in fact they would develop all sorts of weird neuroses thanks to the imperfections of the human shape. Robot chauffeurs might prefer to be built into the car rather than operate it with clumsy human-like limbs; robot spelling checkers would be frustrated at having to fumble with a physical dictionary. A tiny but nice insight ... though sometimes I suspect LocoSpell would like just one hand, to slap our wrists.

Then there's Murray Leinster's little tale 'A Logic Named Joe', which with dazzling good luck did actually manage to predict the desktop home computer – with tele-shopping and on-line database facilities, too. Better still, the plot is actually about the perils of the information explosion, as dutiful video-screens issue detailed technical answers to questions on how to improve one's lifestyle by foolproof robberies, perfect murders, or worse. Rogue data, loose everywhere!

Not bad at all, for 1946.

PCW Plus 68, May 1992

Enquire Within Upon Everything

Lots of computer columns devote themselves to exciting technical tips and advice. It's time I fell into line and gave you the benefit of my hard-won experience, gleaned from long and arduous minutes in public bars. Can I have the first question, please?

• *I'm trying to complete Level III of the Hack's Quest adventure game but am stuck in the Vasty Hall of Rejection Slips without any usable exit. How do I progress?*

This is a very common query. To complete this level of the game you must first have entered the Forbidden Pools in Level II. After persevering for thirty to forty years you should receive the Million Pound Cheque – armed with which, you can obtain some Palm Grease and thus slip by the Ferocious Editor in the Smoke-Filled Room who is blocking your upward route to the Big Time. Also you should stop printing out all your work in single-spaced draft quality at 17 characters per inch.

• *When I ordered the MegaHyperSuper Ultimate Word Thingy program, I expected it to solve all my problems ... but I can't load it. In fact there doesn't seem to be anything on the disk.*

This is a very common query. The software you have acquired is so high-powered that to prevent illicit copying it has been 'read protected' according to access control guidelines laid down by the Federation Against Software Theft. As a magazine reviewer I can tell you that the program is dead good and worth every penny you paid. As a mere customer, unfortunately, you have too low a security clearance actually to use this classy package. A MegaHyperSuper spokesman comments: 'We have to protect ourselves against you thieving sods.' Next?

• *First off, no blinding me with science, right? I don't know anything about these computer technicalities and I mean to keep it that way. You jargon-mongers really do make me sick. Right there on the 'Opening Menu' of your elitist magazine there are all these incomprehensible hardware codes like 0225442244 – I mean, can't you write plain English?*

This is a very common query. It is our telephone number.

• *I have just bought a state-of-the-art IBM 486 system running at 50MHz with a 1200 megabyte hard disk drive, Super VGA colour monitor, CD-ROM reader, fax interface and laser printer.*

This is a very common ... er, what is your query, please?

Nothing, I just wanted to make you jealous.

[Expletive deleted.]

• *I have been told that in order to become a Power User and kick sand in bullies' faces, I should learn the Amstrad CP/M manual. But after just a few paragraphs each time, my brain crashes and I have to be switched off and reloaded from a start-of-day disk. Am I doing something wrong?*

This is a very common query. Stop it at once! It makes you go blind. As is well known in the PCW world, all Amstrad CP/M manuals are infected with a computer virus which inexorably destroys their readers' brain cells, often to the accompaniment of a mocking message about its being Alan Sugar's birthday. Cautious users always practise 'safe CP/M' by covering their hands in cling-film and learning about the operating system only from *PCW Plus*.

• *Here's a useful tip. When ending a paragraph in LocoScript, save yourself a whole lot of tapping at the space bar by pressing the poorly documented key called Return! Look for it above the right Shift key. Not many people know this handy time-saver.*

For this month's Star Letter, you win a bumper pack of 2 self-adhesive disk labels!

• *I am thinking of buying MegaHyperSuper's Ultimate Rainbow Adaptor add-on, which is supposed to upgrade my PCW to a full-colour display for only £3.99 plus VAT. Is the product good value?*

This is a very common query. I've tried out an evaluation copy of the adaptor kit, and must say it's quite ingenious. Just about any colour combination can be quickly achieved, and the transparent felt-pen ink is easily wiped off the screen when you want to 're-customize' it. Yes, with this product MegaHyperSuper (who have just booked a two-year series of multi-page coloured ads in this very magazine) really have a winner on their hands! One tip: if you want, say, all boldfaced words in LocoScript to appear purple, it is important not to type such words in non-purple regions of the screen. Simple enough, but not everyone tumbles to this at once!

• *No doubt I'm just being slow on the uptake, but I keep having trouble with the size of my M: drive and ...*

This is a very common query. Honestly, I don't know why you wallies create such fuss over problems so simple a retarded woodlouse could tackle them. All you need do is make a perfectly ordinary 3.502" hole in the side of the PCW casing using a precision keyhole saw or electric drill attachment. Then, with a standard TR34 hex grommet (non-magnetic), re-align the spode emulator and concatenate the motherboard multiplexor attachment until the third lateral tag flange engages snugly with the plastic lug embossed AK-292664331x-i. Remove all surplus fluff, metal shavings or bitumen, and make good with Polyfilla. Your guarantee should now be successfully invalidated. I forgot to say you should have switched off first ... oops, too late now.

• *Why spend money on ribbons all the time? Here's my tip: put a sheet of carbon paper over each fresh sheet you load into your PCW printer. You'll get perfect – well, almost perfect – print and will never again need a new ribbon!*

Er ...

Come to think of it, why waste money on expensive carbon paper? If you make regular credit card purchases, shop assistants will be delighted to give you the left-over carbon slips from the little chit thingy. With just a bit of time and patience these can be gummed together into perfectly serviceable A4 sheets!

This is a very common complaint. You have what medical men call Ebenezer Scrooge Syndrome. Unless you can pull yourself together and treat your poor PCW more generously, three dread spirits are likely to appear to you next Christmas ... and the horrible part is that they will all take the form of Alan Sugar.

• *I really appreciate the personal touch and amazing erudition you bring to the Technical Tips page. No question seems to stump you! How on earth do you keep it up?*

This is a very common query. The answer is, quite simply, ILLEGAL INPUT ????????.$$$ FATAL PROGRAM ERROR IN QUERY MODULE 'LANGFORD' – PROCESSOR STACK FAILURE – MASSIVE BDOS COLLAPSE – Retry, Ignore or Cancel?

PCW Plus 70, July 1992

Eureka! The Wheel!

Re-inventing the wheel in software is supposedly a thing to avoid. What a waste of time, when some suitable utility program is sure to be available. But that's not how it works out. Instead ...

Well, I'm thinking, this is it. Software supremo Langford exposed as a recession-hit cheapskate. Years ago I said airily in print that I could handle any PCW disk format, and now this guy wants stuff copied to a perfectly ordinary PCW disk. Only it's 3½" and I haven't got one yet. Nor do I want to risk upgrading the old 8512 when I regularly use both its drives.

Really we should have the 3"-drive 9512 converted.

Except I tripped over its printer cable and triggered the Amstrad Surprise Circuit which, when you wiggle the plug, blows out the RAM. Can't afford repairs just now. (Can't afford a 3½" drive upgrade either, actually.)

But the pride of the Langfords demands that I produce a 3½"-format disk this week. No, I have the disk. The chap sent a blank, formatted one. If I could *copy* the wanted files to it ...

My secret programmer's instinct, based on the hard-gleaned wisdom of years, suggests that it would not be a good idea to saw one of the 3" drive slots a bit wider.

Pause to pace the office, kicking final demands out of sight. All I need is a 3½" drive. In fact I am surrounded by 3½" drives, on obsolete Apricot computers, on the Amstrad PPC with the unreadable duckpond-murk screen, on the IBM ...

Wait a minute. The IBM is cable-linked to the PCW for file transfer. Hoping against hope, I stick the PCW 3½" disk in the IBM drive and type DIRectory. 'General failure reading drive B,' is the sneering screen message, IBM's way of saying 'You're not catching me like that, squire.'

Thinks. You can get software to read and write IBM standard-density disks in a 3½" PCW. They're the same disks as PCW ones: double-sided, 720k capacity, even *the same shape*. Just a different format. Ergo, with the proper software I should be able to read and write PCW disks in the IBM drive. Even format them.

Marvel at the brilliance of my deductions. Consider running naked through the streets shouting 'Eureka!', but Britain perhaps less tolerant of genius than ancient Greece.

They say: never waste time re-inventing the wheel. If a piece of software seems useful it surely exists, and something as small and handy as a CP/M-format disk thingy for IBMs is bound to be shareware or public domain. Quick circular to PD libraries and disk transfer specialists using the office fax. Wait.

Wait.

To give them credit, West of Britain Business Services (only) do reply. Wonderful people. Unfortunately they brush aside my enquiry and suggest a 3½" drive upgrade plus Moonstone software to read IBM disks in the PCW, which was not the point.

Now what? I know roughly how CP/M disks are organized ... no harm, surely, in writing a few lines of program to have a peep at the directory structure? Except I only have a blank PCW 3½" disk. *That* is soon corrected: a credit card order for a random public-domain something to Advantage, who, bless 'em, deliver by return of post. Meanwhile, I prepare a rough program. This way madness lies. Believe me.

How does it go? When they're formatted these 3½" disks are laid out in 80 tracks (concentric rings on the magnetic surface), each track divided into 9 sectors, each sector having its counterpart on the other side of the double-sided disk. A sector, by immemorial tradition, holds 512 bytes or text characters. Multiply up and you get a disk capacity of 737280 bytes, but part of that isn't for you: it's reserved for CP/M or LocoScript 'housekeeping' information. Like the directory.

Squint at the public domain bargain with hasty home-made software. First thing on the disk is sector zero, the 'boot' sector, with arcane machine-code instructions for loading a CP/M or LocoScript start-of-day file. Sectors 9 to 24 contain the disk directory.

(These, gibbering readers, are figures for 3½" PCW disks.)

Insanity mounts. A directory entry takes up 32 bytes, so there can be 16 per directory sector, 256 altogether. In each entry, the first byte says which group the file is in; the next 11 give its name (omitting the dot); next comes the one-byte 'extent' (one entry can catalogue only 16k of a file, so longer files are divided into 'extents' with the same filename but numbered 0, 1, 2 and so on); then two bytes whose purpose I forget (I think it's optional date/time stamping); then a byte recording the number of 128-byte chunks of the file in the current 'extent'; and finally up to 8 two-byte numbers, codes for the 2k blocks of actual disk storage used by the current 'extent' of the file.

To make life more interesting, the first 2k block has code number 4, indicating to the cognoscenti that that bit of storage begins at sector 25 of the disk, just after the directory itself....

Who thought up all this stuff?

So. To copy a file to this disk without benefit of CP/M or LocoScript, your IBM program merely has to *(a)* divide the file into 16k 'extents'; *(b)* find a free or deleted directory entry for each – I forgot to say the group number is set to 229 when an entry is deleted; *(c)* read all the block code numbers which are in use, crossing them off an imaginary list to find if there are enough free blocks to hold your file; *(d)* copy the file, a bit here and a bit there, to these vacant properties; *(e)* overwrite empty or deleted directory entries with new ones that will guide other programs to where the bits of your file may be found, in the right order.

Clever little sods, these programmers.

Everything after that is a blur. One pal always warns that 'software is a *disease*'; another likens it to the rapture of the deep. Hours pass like minutes, days like hours. Instant copier program working perfectly at only the 5,271,009th correction. Files copied from 3" PCW disk via cable to IBM, via new program to CP/M 3½" disk in IBM drive, read back again, checked, double-checked ... Honour of Langfords saved. Customer happy. Kindly hands remove me from the keyboard, blurred figures in white coats say 'Drink this.'

I just know that tomorrow or the day after, someone is going to send me a public domain copier that does *exactly the same thing*, but until then I intend to feel smug. Shattered, but smug.

PCW Plus 72, September 1992

How Not to Write

Writing can be a terrible curse. I know tormented souls who are unable to stop (there is even a Latin medical term for this: *scribendi cacoethes*) and who travel everywhere with thick wads of continuous print-out which they will relentlessly show to anyone who offers them an opening by mentioning books, computers, paper, beer, the weather, etc.

Luckily for these unfortunates, using a PCW or other word processing system offers a vista of new alternatives to the fearful writing compulsion! Not only is a computer display several times more efficient than a piece of paper in producing that welcome headache which forces you to go and lie down for a bit, but the machine also offers countless distractions. Even while avoiding all games and keeping everything relevant to writing, one can emerge from a long busy day in the green deeps of the computer screen without having written a single word of that great novel (which, after all, you'd only have to mail out in hope of interestingly phrased rejection slips). Here are a few ways.

Distraction 1: Having a Good Whinge. All computer firms have signed the Customer Annoyance Pact, agreeing that at random intervals they will spring irritating surprises such as discontinuing your favourite program, computer or disk size. (The new IBM jape is software on CD-ROM, with a hidden cost of £300 for a drive that'll read it.) Writing to them about their cruel decisions can offer the same high exhilaration as smashing your head repeatedly into a polished slab of teak.

For example, a thrilling correspondence with dear old Amstrad Customer Services recently fell into my hands:

'Is it still possible to get decent replacement ribbons for the Amstrad LQ5000di printer? By "decent" I mean "PT NO. Z70360", which came with the printer, rather than "soft no. 50029" which seems to be the only thing available in my local office equipment shops, and which appears, on internal inspection of both, to have only a quarter of the ribbon (if that) of the Z70360. This means, of course, that it needs to be re-inked four times as often, and will eventually wear out more quickly and have to be replaced altogether....

'The 50029 case also appears to be more cheaply constructed than the Z70360 case. After just one re-inking I found that the ribbon tended to slip out at the end of the left-hand "horn"; when I took it apart to repair it, several of the studs which hold the case together broke. It's now held together with masking tape. Also, the Z-shaped little plastic hooks which hold the cartridge in place in the printer snap off very easily on the 50029, a problem I have not had with the Z70360. I have just had to buy a second spare ribbon (another 50029; that's all I can find), only a couple of months after buying the first.

'Is it possible to me to get hold of Z70360's (or genuine equivalent) from you or another outlet?

Back came a helpful reply from the Amstrad Information Centre, explaining that 'the supplier of ribbons has been changed and therefore only ribbons with "soft no 50029" are now available' ... also suggesting that the complainer buy lots from Amstrad Spares Direct Sales at once.

Our man wrote again, saying among many other things 'A company of your size and market position is quite capable of telling a supplier what is required in a product; changing suppliers is not a valid reason for substantially reducing the quality. The customer, if necessary, would be prepared to pay a little extra for a ribbon which lasts four times as long.'

An even more helpful reply from Amstrad said soothingly that they weren't interested as 'complaints of this nature are very few'.

The complainer's blood is up ('I intend to win'); the correspondence continues; and, several letters later, imagine how much keyboard time has been spent at this creative work rather than boring old journalism.

Distraction 2: Make It a Work of Art. My heart always sinks when I receive a letter plastered all over with the unmistakable signs of someone who has just discovered Desk-Top Publishing. Ten or twenty garish fonts, you know, and little boxes with shadow effects surrounding the address, the date and the signature, and blurry patches of clip-art everywhere ... a digital dog's breakfast which took only about twenty times as long to produce as a LocoScripted letter.

DTP is like word processing in one important way. People don't as a rule assume they can load a drawing program and instantly create great Art, or that the spreadsheet will at once teach them accountancy – but everyone 'knows' he or she can already write and lay out a page professionally. Goodness knows why anyone ever paid good money to typographic designers or layout artists.

Clip art is the most insidious DTP snare of all. The first stage of addiction involves buying countless disks of the stuff, megabytes and megabytes, and spending bleary hours watching each image appear v-e-r-y s-l-o-w-l-y on the screen, hoping that *this* will be the perfect illustration at last.... Well, perhaps the next one.

(Is copyright *different* in the clip art world, by the way? I have seen 'public domain' disks containing scanned versions of Escher woodcuts and lithographs whose copyright I thought was jealously guarded by the Escher Foundation at The Hague. I have seen a PCW-produced science fiction fanzine full of cartoons lifted straight from recent national newspapers. But my lips are sealed.)

I now have second-stage clip art addiction thanks to buying a scanner. Once I used to photocopy images and just paste them on to the print-out: now science has found a slower way to do it.

In fact you don't spend that much time actually scanning things; you become a relentless scavenger, turning over the pages of every book you own in hope of some suitable image to go with what you're writing. 'No, that one's too small ... not quite relevant ... too big ... too much fine detail ... is this chap da Vinci out of copyright?' Then you stumble on the one perfect illustration, and the publishers have printed it in pale blue over a page of text. Eyes bulge, fingers throb, existence slows to a glacial pace, and unfortunately I own 20,000 books.

Distraction 3: Waving Large Magnets Over Your Clip Art Disks. [I think we could rather usefully stop there – Ed.]

PCW Plus 74, November 1992

Future Shock

I suppose – I hope – it must have been a brainstorm brought on by my recent, masochistic task of reading, on disk, the whole 1,200,000 words of the coming new edition of the *Science Fiction Encyclopaedia*. (Out in 1993 if all goes well.) But perhaps one of those strange SF warp thingies did indeed open up in the fabric of time and space ... and instead of its traditional functions of causing crop circles, UFO sightings and missing address marks, it gave me a fabulous vision of the future.

In this utopian vision, gentle reader, I saw an issue of *PCW Plus* from several years or decades hence. It was all-electronic, of course, presented on the A3-format high-resolution screen of a PCW 1048576+. The hardware apparently used a real-time optical tracking system, since the cursor automatically moved to wherever your eyes focused on the screen, and you could select menu options by blinking. In fact one of the advertisers took unscrupulous advantage of this with a colour picture of a deeply sexy lady – whenever you blinked at her (and there was good reason), the screen immediately lit up with 'Order accepted and transmitted to Ansible Information: the cost has now been deducted from your credit account, including postage and VAT at 27.5%'. But I digress.

The lead article in this *PCW Plus* was (will be? will have been?) all about the release of LocoScript 48.1, beginning with a familiar-sounding explanation of how ordinary people just don't need a 100-terabyte computer running holographic virtual reality systems in order to do word processing, and explaining how the PCW's simple gigabyte hard disk, 16,384 screen colours and twin CD-ROM drives were still an ideal budget solution at only four and a half thousand New Pounds, or barely two-thirds the price of a brand-new scrotty ... I never did find out what a scrotty was. LocoScript 48.1 came packaged free with the current PCWs (except the bottom-of-the-range model, which still has only LocoScript 1) and offered some remarkably powerful new features.

For example, expert users can apparently set up the word processor to generate entire novels with only seven to ten keystrokes. The built-in style checker will mutter a discreet 'Tut-tut' through the PCW's hi-fi quadraphonic speaker system whenever you write an unclear sentence or use a sexist pronoun, and can also offer advice on your personal and financial problems. Other useful features are the Limerick Formatter and the Plagiarism Checker. It certainly sounds like a wonderful program, although for technical reasons it has still proved impossible to include a word counter that runs without loading all 320 megabytes of LocoSpell.

Turning to the adverts, I found plenty of new products. The independent release of the month was of course the ten-volume *Easy Guide To Loco 48.1* ('Why struggle through all twenty-one volumes of the complete manual? We tell you what you need to know'), hotly rivalled by the four-volume *Compact Guide to the Easy Guide*, the two-volume ...

The Flipper program seemed to have kept pace with the development of the PCW, too, and according to an aside in this future magazine Flipper is now able to make the computer multi-task as a simultaneous word processor, spreadsheet, database, VCR, coffee maker, scrotty simulator and small electric automobile (add-on battery pack or very long power cable recommended). And could it be true that as claimed in *News Plus* in the same issue, *The Sun 3-D Holo-Supplement* was now produced entirely on a single PCW running the new release of Mini Office Professional?

Some 'pages' looked more familiar than others. The *Postscript* feature was comfortingly full of people complaining about incomprehensible manuals, overpriced software, and defective hardware ('The PCW power supply is ludicrously inadequate to run a standard add-on 48-user network system'). A special guest article brought tears of nostalgia to my eyes with its list of amusing and unexpected substitute words suggested by LocoSpell 36.2 (now incorporating the complete *Oxford English Dictionary*). *Tipoffs* provided much wise advice on CP/M 7.0 filenames ('only 64 characters are allowed before the dot, and 16 after'), how to learn the use of PIP in mere days and how not to spill steaming, heavily sugared coffee into the keyboard. But I couldn't understand the *Competition* at all: it seemed to be all topical jokes about *Alien XI Meets Batman*, Prime Minister Sutch and scrotties.

Listings had become a lot bigger owing to increased storage space and fractal compression techniques ... this issue's offerings included the Los Angeles *Yellow Pages*, the complete out-of-copyright works of Rudyard Kipling and – presented for its military-historical interest – all the millions of lines of programming written 'long ago' for the US Strategic Defence Initiative. There was also a grandmaster chess program that ran in ten tightly-coded lines of BASIC ... I guessed that BASIC must have been upgraded a bit too.

As I scanned these wonders of an uncertainly distant future (I just couldn't seem to find the cover screen with the issue date), I stumbled on an electronic page of peculiar interest to myself. It was headed *LANGFORD*, but there was only one small paragraph there, edged in black. As far as I can remember, it read: 'David Langford is unwell. As readers will know, our long-time columnist had an unfortunate accident when an infodisk he was cortex-uploading turned out to be contaminated with the WOMBLE virus, which erased his entire brain. These things will happen! Dave's Disk Doctor Service has been using its neuroassembler toolkit and quantum probes in an attempt to retrieve the shattered remnants of the Langford intellect. We now hope to have him up and running in PCW simulation in time for Christmas ...'

Suddenly it became very important to me to know the date of this issue of *PCW Plus*. I flipped faster and faster back through the electronic pages and had spotted the words OPENING MENU when the PCW hypertext display blurred ... dissolved ... was replaced by the full-colour image of a grim-jawed man looking like Arnold Schwarzenegger but not so soft and cuddly. 'This is a recorded message from the Federation Against Software Theft,' he boomed. 'Automatic laser scanning of your retinal patterns shows that you are not an authorized *PCW Plus* subscriber or purchaser. The lethal copy-protection system now going into operation is authorized by the Computers (Misuse) Act of 1997 and is in no way a violation of your EC human rights under the Treaty of ...'

Then everything went black.

PCW Plus 76, January 1993

Golden Phrases

One of my secret addictions is buying and skimming tatty old books on the secrets of professional writing ... especially when they're so out of date that the chapters on tax and libel are mostly about how to avoid Morton's Fork and the Star Chamber. In fact my latest scheme for getting rich is to transcribe some of these out-of-copyright wonders into vast LocoScript documents that are bound to do something for your writing, possibly something terminal.

For example, my 1909 *How To Write Letters That Win* is full of useful examples that every PCW user will surely want to have handy on a disc for the next time they do business by mail – in particular when selling wooden buggies, bespoke suits, St Andreasberg Roller Canaries or (and I quote) 'patent-lined, double-rimmed, rust-proof, excelsior gas burners'.

A particularly timeless moment in this anonymously written masterpiece comes when you get to the sample letter that begins 'You wouldn't think of throwing away your fountain pen simply because the ink is exhausted' ... and goes into enthusiastic detail about re-inking old typewriter ribbons via a 'special process'. I must look up the small ads for PCW ribbon re-inking services to see if any of them say 'Est. 1909'.

I also like the model letter to a customer who has dared to complain: 'Dear Sir: Your eyesight must be going back on you. The paper you ordered is certainly identically the same stock as the sample you named. Take it to the window and look again.' Alas, this turns out to be a Bad Example, and the book goes on to suggest a Correct Response which grovels so nauseatingly that I can't bring myself to quote it.

The book in my collection that truly cries out to be put on disc is *Fifteen*

Thousand Useful Phrases by Grenville Kleiser (1917), which is rather beautifully divided into Useful Phrases ('comatose state'), Significant Phrases ('unseemly and insufferable'), Felicitous Phrases ('faded, dusty and unread'), Impressive Phrases ('giddy, fickle, flighty and thoughtless'), Prepositional Phrases ('zone of delusion'), and so on. Whenever I open the thing I find myself caught and hypnotized by the magnificent sections called Literary Expressions and Striking Similes. With this book, stories practically write themselves! You are about to be very unjust to me. Although I haven't tampered with a single golden word of it, you will shortly believe I am making all this up.

All right, our story needs characters. Here is the description of a chap: 'A broad, complacent, admiring imbecility breathed from his nose and lips.' Nor is his conversation up to much: 'A dire monotony of bookish idiom. He clatters like a windmill. He spoke with a uniformity of emphasis that made his words stand out like raised type for the blind.' He dresses well but doesn't wash enough: 'Wrapt in his odorous and many-coloured robe ...' Overall he is rather 'Like some suppressed and hideous thought which flits athwart our musings, but can find no rest within a pure and gentle mind.' But at least he does have 'A large, rich, copious human endowment.'

Mr Kleiser's inexhaustible sourcebook quickly suggests a female character to counterpart the hero. 'She exuded a faint and intoxicating perfume of womanliness, like a crushed herb.' She comes from 'A puissant and brilliant family' and has 'An avidity that bespoke at once the restlessness and the genius of her mind.' But alas, 'Every curve of her features seemed to express a fine arrogant acrimony and harsh truculence,' and moreover 'She flounders like a huge conger-eel in an ocean of dingy morality.'

Of course they meet, and the plot develops: 'A quick flame leaped in his eyes. A queer, uncomfortable perplexity began to invade her. A river of shame swept over him. A shiver of apprehension crisped her skin. A new trouble was dawning on his thickening mental horizon. The music of her presence was singing a swift melody in his blood. All around them like a forest swept the deep and empurpled masses of her tangled hair.' She is duly 'Flushed with a suffusion that crimsoned her whole countenance.' Soon they both 'Clutch at the very heart of the usurping mediocrity,' and so would you.

The scene shifts from place to place: one moment they are 'Covered with vegetation in wild luxuriance,' the next 'The landscape ran, laughing, downhill to the sea' where 'The murmur of the surf boomed in melancholy mockery.' Next paragraph 'The hills were clad with rose and amethyst,' and soon both characters are 'Grazing through a circulating library as contentedly as cattle in a fresh meadow.'

I suppose the sexy bits have to come here (suitably 'Clothed with the witchery of fiction'). 'A quiver of resistance ran through her,' and he is 'Beside himself in an ecstasy of pleasure.' So naturally 'He treads the primrose path of dalliance. He smote her quickening sensibilities.' Next comes a 'Delicious throng of sensations' and the pair are 'Fatally and indissolubly united. It was sheer, exuberant, instinctive, unreasoning, careless joy. Volcanic upheavings of imprisoned passions.' In short, there is 'A lapse from the well-ordered decencies of civilization.'

Subsequently finding that 'Love had like the canker-worm consumed her early prime' and feeling 'A glacial pang of pain like the stab of a dagger of ice frozen from a poisoned well,' our lady soon changes her mind about the relationship. 'Her scarlet lip curled cruelly' and she sneers: 'I yielded to the ingratiating mood of the day. I capitulated by inadvertence. Banish such thoughts.' He complains: 'You gave me such chill embraces as the snow-covered heights receive from clouds. So my spirit beat itself like a caged bird against its prison bars in vain.' He says this because 'He could detect the hollow ring of fundamental nothingness. He was giving his youth away by handfuls. His reputation had withered. He was quaking on the precipice of a bad bilious attack.'

Thus, 'Slowly, unnoted, like the creeping rust that spreads insidious, had estrangement come. They became increasingly turbid and phantasmagorical. A remarkable fusion of morality and art.' Let's just leave them there, shall we? – 'Sunk in a phraseological quagmire.'

Perhaps our favourite software firm could reissue this wondrous book as 'LocoPhrase', guaranteed to brighten up anyone's writing.

PCW Plus 78, March 1993

Those Crazy Ideas

It's an old question: where do writers get their ideas? The word processor sits before you with its forbiddingly blank screen, demanding words to process ... and they have to come from somewhere. Rather than get all metaphysical about the ultimate origins of words and ideas ('minutes after the Big Bang, the superheated semantic flux began to condense into the universe's first pronouns and indefinite articles'), let's survey the practicalities. What makes ideas come, and what stops the flow? Where can you reliably find one? In no particular order....

Schenectady. This is a standard answer given by American SF writers tired of being asked, 'Gee, where do you get those crazy ideas of yours?' Saying with weary patience, 'a mail order service in Schenectady,' is supposed to shut up enquirers. In fact they usually come back with, 'Hey, can you give me the address?'.

Dreams. Some writers swear by dreams as a source of images, a lucky dip into the murky waters of the mind. First you need to be a vivid dreamer; then you train yourself to write down all interesting dreams the moment you wake. This applies especially to dreams so striking that you can't *possibly* forget them. (You will.) I have managed to dream entire story plots in my time, but they never seem quite convincing in the cold light of morning. Single nightmare images are often more durable, and I've used a few in horror stories. Bob Shaw's best-ever nightmare involved falling into a vat of tiny ball bearings and feeling them clicking inexorably against his back teeth on the way to his stomach. You'll find this incident in his SF novel *One Million Tomorrows*.

Noise level. Some writers function better to loud background music, or even VERY LOUD background music. Rather them than me.

Diet. Being too well-fed can have a deadening effect. Jerome K. Jerome of *Three Men in a Boat* fame noted that a sufficiency of hot muffins would always leave him 'dull and soulless, like a beast of the field – a brainless animal with listless eye'. The moderately obscure SF writer John T. Phillifent (alias John Rackham) claimed that he could give himself creatively useful nightmares and hallucinations by avoiding vitamin B1. Cut out B1 sources like cereals, liver, bacon, eggs or yeast in any form, he suggested, and develop a useful deficiency disease! *Please note: this is not recommended to anyone.*

Self-delusion. The yawning computer screen, the aching sense of having to write something and write it right, can paralyse thought. Tell yourself you're only joking, playing at writing, and type a bit of free-association prose – anything to break up that awful emptiness. The humorist Patrick Campbell, when desperate to get a comic essay started, said he used to sneak up on the keyboard and type: '"Well, what about Harold Wilson now?" he said.' Once he'd doodled a few more lines and got some kind of conversation going, he could go back and substitute an opening to match the emerging punchline.

Drugs. Caffeine fuels most writers: quart after quart of coffee, pot after pot of tea. Alcohol is a mistake, despite those tales of Nordics who held drunken meetings so ideas could flow freely (followed by a sober and hungover meeting with power of veto, for caution's sake). The trouble is if you let yourself believe that a single small drink will relax you and release the creative flow, this slides rapidly into the twin delusions that (a) you always need that drink even when starting first thing in the morning, and (b) if one dose doesn't do the trick, a second will. I don't need to say anything about serious drugs, do I? You're not daft. Lord Dunsany wrote a fantasy about a 'Hashish Man' who found the secret of the universe in a dope dream ... and he got a fan letter from famous loony occultist Aleister Crowley ('The Great Beast'), saying it was evident from the story that Dunsany had never tried hashish. He hadn't, but his imaginative version was better than all too much limp fiction written by the later hippies who thought they *could* find ideas that way.

Serendipity. You can't organize luck, but you can give it every chance to turn up. Potter at random among your books, especially ones unrelated to what you're writing. (If you don't have a vast number of weirdly assorted books, are you serious about being a writer?) I was stuck trying to produce a 'creepy' story and distracted myself by skimming through miscellaneous non-fiction: suddenly, in a popular maths book, I found a passage about the superstitious horror of Pythagoras when he first stumbled on the irrational numbers. One thought led to another, and the story was soon roaring ahead.

Real Life ... is a dangerous source. That incredibly funny thing so-and-so said at the office may capsize your writing thanks to the sheer weight of background and build-up needed before the point comes properly across. The same goes for the amusing incident at the supermarket checkout. Never let real life into fiction without first creatively overhauling it. This is also true of non-fiction; there are many different ways of telling any given truth (and the most stark and objective-seeming version may have a more slanted effect than some elegant reworking).

Plotto. Over the years, countless mechanical plot generators have been marketed: books, packs of cards, computer programs. None of these systems is going to write a story for you. Could it spark one off? This depends on whether you're receptive to such nudging. Some people write down all their ideas and concepts on bits of paper and shuffle them in hope of inspired juxtapositions. If feeling high-tech, you can do the same in a BASIC program of random choices (see RND and RANDOMIZE in the manual). Or buy Ansible Information's legendary *[No plugging yourself this month, Langford – Ed.]*

All these ways of jogging creativity, all these little enemas for the Muse (in critic Nick Lowe's regrettable phrase), are mere preliminaries to the titanic struggle of naked brain against blank PCW screen. Sometimes I think it's the most absorbing game on earth. Other times I just want to scream. Yes, *screaming* is another popular technique employed by many authors ...

PCW Plus issue 80, May 1993

News on the March

This magazine has featured many a spine-tingling article about producing newsletters on your PCW ... but nothing as delirious as the one I recently edited. It appeared at the frantic pace of two issues a day, to be read by close on 1,000 people, and no luxuries like photocopiers or offset-litho machines were permitted. Insanity!

The setting was Helicon, the British national science fiction convention, held in St Helier, Jersey, over Easter 1993. (This was combined with the Europe-wide 'Eurocon', leading to a rich, multi-national mix – I don't suppose there had *ever* been as many as 52 Romanians in Jersey before.) A twice-daily newsletter is traditional at these events, and traditionally, each year, some gullible idiot is persuaded to edit the thing.

Reader, that gullible idiot was I.

Luckily I had a supporting cast of dozens, including the Technical Editor of the brand-new edition of *The Encyclopaedia of Science Fiction* (plug, plug). I also managed to devise a suitable name for the newsletter, *Heliograph*, and to get some advance material on disk before the convention began – notable SF birthdays, anniversaries, famous authors' reminiscences of past conventions, topical quiz questions, and so on.

For example, my twisted researches produced a 'news' item on a tricentenary of vital interest to us all: 'In 1693 Gottfried Wilhelm Leibniz of calculus fame invented the first mechanical calculator that could multiply and divide, thus heralding an exciting new era of arguments over the restaurant bill. ("Fie on you and your Engine, fir, I had only a fmall falad and a Pepfi.")' Keeping it entertaining, even with such arrant padding, was important: the funny bits were the sugar coating which persuaded convention members to gulp down all the worthy announcements and urgent notices of programme changes.

There was nothing particularly unusual about the medley of computers we used, but as each page rolled out of the printer the technology slipped several decades backwards in time. The low-budget printing system was a marvel of industrial archaeology.

First came the electronic stencil cutter. You wrap your master print-out around this long rotating drum, and next to it is wrapped a vinyl 'electrostencil'. There is great fiddling with knobs, adjustment of light-bulb brightness and nervous checking of meters. Then the BIG BUTTON is pressed and the whole contraption thrums into life, with a photocell tracking along the spinning print-out as though it were an old-time phonograph cylinder. A stylus needle moves in step along the electrostencil, reproducing the printed pattern of black and white by literally burning through the stencil with an electric spark. Clouds of ozone and carcinogenic fumes billow out....

I don't think the 'newsroom' where I more or less lived for a week was a very healthy place. Whenever the electrostenciller was going, layers of black dust rapidly collected on the computer screens, stuck in place by static. And it took a quarter of an hour to cut each stencil.

Then, assuming that everything had worked, the stencil was mounted on the drum of an ancient Gestetner duplicator, and thick, gooey ink began to spurt in all directions. Older readers will know that the principle of the duplicator is incredibly simple: wherever there are holes in the stencil, ink seeps through to mark the paper which is being cranked through the machine at desperate speed. A certain wastage is caused when the ink comes through too fast, or not fast enough, or runs out, or glues the current sheet of paper mercilessly in place, or ...

I think I might demand a boring old photocopier if I ever volunteer for this again. There also exists a wonderful (though huge and expensive) machine made by A.B. Dick which has an electrostencil cutter and duplicator bundled together in one big case: you shove in your master sheet and clean copies automatically spew forth by the thousand. But the secret advantage of *Heliograph*'s Industrial Revolution set-up was one which every PCW owner will appreciate: it was dirt cheap.

And then, when the dud sheets had been shovelled aside and the ink had dried almost enough not to smudge, the new edition was triumphantly distributed to waiting SF fans, while the editor heaved a sigh and started trying to think of jokes for the edition after that.

If you ever produce a newsletter on the run like this, here are some bits of hard-won advice:

One person (the editor) must take responsibility for final 'house style' editing and printing the master copy. With a motley crew of volunteers, there is no time to tell everyone how you want things done.

Rule 42, as we know from *The Hunting of the Snark*, is 'No one shall speak to the Man at the Helm.' No matter what chaos is raging all around, the editor must be left in peace to edit as the deadline creeps near.

Everyone finds it madly irritating to have people look over their shoulders commenting on typos. Try to arrange the desks or tables so it's difficult or impossible to 'overlook' the monitors.

You need at least two computers, so that people who burst in with shock announcements can be told, 'Don't ask us to memorize it: type it in.' Bits of scrap paper for scribbling down odd notes will be spontaneously generated by the printing process.

Exiting the word processor and running a communications program on two computers is far too much hassle. Passing disks across the table is the sane way to move text to the master version.

All the following will be desperately needed if you didn't bring them: paper tissues, Blu-Tack, Tipp-Ex (this was the one I forgot, whereupon our borrowed laser developed a flaw on the drum and left ugly black marks in every margin), craft knife, aspirin, patience, more patience.

Be nice to your volunteers.

Oh: for details of the 1994 British Easter SF convention, write to *[but that address is long out of date – point your web browser to http://links.ansible.co.uk/ for a current list]*.

Finally ... one favourite news item from *Heliograph* was the information that SF authors Brian Aldiss, Harry Harrison and Anne McCaffrey had all joined the KGB.

With capitalistic flair, some visiting Russians were making a very nice thing out of selling obsolete KGB credentials to Western SF fans. Think of that.

PCW Plus 82, July 1993

• *At this point my column was dropped: a sharp reduction in the amount of editorial (i.e. non-advertising) matter in* PCW Plus *indicated the magazine's shakiness at this time. But the hideous Langford presence returned on a bimonthly basis in 1994:*

Caught in the Net

You've probably puzzled over many newspaper articles on electronic mail, computer nets and the Internet. Perhaps friends have started quoting irritating 'net addresses' with @ signs swarming in them like horrid bacteria. (Me, I'm ansible@cix.co.uk.) Very recently British journalists became 'net-aware', largely because more newspapers and magazines are accepting articles via e-mail: journalists, being lazy, have delightedly realized that they needn't print stuff out or use that old-fashioned fax machine....

I can't resist novelties, and have duly experimented with the net. As usual, the real thing has quite a different flavour from the reports of newspaper hacks trying to make it sound exciting by their standards. At risk of boring you rigid, let's correct some current misinformation.

• 'You can't use computer nets with a PCW.' This is true in the same sense as 'You can't do desktop publishing on a PCW.' Yes, an IBM is more convenient, but patience and a bicycle can get you there as effectively as a car. Stumbling blocks for PCW users are that you need a serial interface and a modem, and that their extra cost makes communications software look uninviting: 'I have to buy these boxes before your wretched program will work?' Cheaper than an IBM, still.

• 'It's too expensive for home users.' Britain has two economical routes into the Internet. CIX – Compulink Information eXchange, with over 10,000 subscribers – is a full-fledged service provider costing a minimum £6.25 plus VAT monthly, being an advance payment against £2.40/hour off-peak connection charge. I can just about stay within the £6.25, except sometimes. For heavy users, the bare-bones service called Demon offers unlimited Internet access for a flat £10 plus VAT monthly. Either way, much of my large overseas correspondence goes via Internet for a fraction of a penny per message, rather than air mail at 41p.

• 'The phone bills are ruinous!' Personally I stick to cheap-rate periods and weekends. But the pleasant surprise here is OLR or Off-Line Reader software. Instead of clocking up connection charges while typing e-mail messages and reading those received from the net, you let this software transmit and receive everything in one automatic burst. Reading messages and writing replies can wait until you've safely hung up. I hardly dared believe that anyone had written such software for PCWs, but all knowledge is on the net: my CIX query brought responses from proud PCW users. I boggled at the chap who was connecting to CIX via a 9512, from Brunei! Margolis & Co. offer a communications program called COMM+ which comes with CIX offline-reader facilities; they also produce PCWfax, whose special serial interface supposedly allows data transfer to and from the net faster than my IBM will do it....

• 'I've read articles about Internet (what *is* it? Why do some people say Usenet?) and it sounds full of incomprehensible jargon.' Right. The world is full of conferencing systems like CIX or the American biggie CompuServe: a combination of virtual-reality chatlines, mailboxes and libraries. Internet is a system of global connections, joining thousands of nets via what SF writers call cyberspace: from CIX, I can post an electronic letter which goes through the Internet link to any net address on the planet. Usenet is a gigantic bulletin board that comes with the Internet territory: a flood of information and chat running to tens of megabytes a day. I couldn't possibly read all that, and just stay connected to a few SF topics ('newsgroups') that interest me – these have only about 100,000 readers as opposed to Internet's overall total of twenty or thirty million. The Usenet technical newsgroups feature a lot of jargon, but other subjects are discussed in plain(ish) English – and there's a newsgroup for almost any subject you can imagine. Including Terry Pratchett. *[Yes, alt.fan.pratchett still continues.]*

• 'What about those stupid abbreviations and smiley faces?' They're not compulsory: I dislike them and don't use them. Some net-addicts are computer nerds with no sense of humour who can't recognize a joke (let alone difficult stuff like sarcasm) unless warned by a little sideways smiley face ... :-). And it's true that writing something funny on CIX will cause at least one acronymic response of 'ROFL!', for Rolls On Floor Laughing. After studying this, a friend of mine moodily suggested 'ROFV': Rolls On Floor Vomiting. News reports always seize on these cranky and eccentric aspects for the sake of an amusing story, and play them up at the expense of more ordinary (and more interesting) electronic chatter.

• 'I hear it's full of filth!' That's the trouble with free, uncensored communication: a very few people use it to send digitized naughty pictures. The way politicians react seems comparable to demanding a ban on all magazines, including *PCW Plus*, because some feature the kind of recreational exercise enjoyed by Cabinet Ministers. (I confess to wondering what exactly is in the Usenet 'alt' or 'alternative' newsgroups called alt.sex, alt.sex.bondage, and worse – but CIX tactfully screens these out.)

If *PCW Plus* readers are interested I'll write again about the net; it's far more fascinating than its current public image. I won't forget the thrill of requesting some free software through Internet and realizing that it was coming straight from a friendly computer at Nanyang Technological University in Singapore – and no international phone charge either....

PCW Plus 92, May 1994

In the Beginning

'Nothing much was happening.' 'He felt like a nice cup of tea.' 'It was quite dark.' 'Statistical analysis indicated a slight upward market trend at the 0.1% significance level, she thought excitedly.'

No, as opening sentences these are duds. As so many PCW users are hopeful writers, I keep returning to the mysteries of writing here – and once revealed the dark secret that most neatly printed submissions aren't read beyond page one. When your editor is jaded, decisions can be based on the first few lines; if there's nothing there to hold attention ... tough luck.

The temptation is to throw all one's efforts into an irresistible opening sentence that grabs readers by the, er, by very tender parts and forces them to read on:

> As the reactor meltdown alarms shrieked OVERLOAD and hordes of rabies-infested mutant wombats closed in, Felicity wrested the stolen Crown Jewels from the snarling Archbishop and hoped her zeppelin's fading buoyancy could drift beyond range of the gunfire from Stonehenge to reach higher ground before the tidal wave hit ...

At this point, groaning at the chapters of flashback required to justify such excitement, the editor may take an aspirin instead of reading on.

What makes a good opening line? Think which fictional first lines you can remember (even approximately) without thumbing through books. I tried this thought experiment and betrayed a misspent youth by recalling lots of SF.

'I always get the shakes before a drop.' Robert Heinlein was famous for zappy openings: here in *Starship Troopers* the narrator is about to be dumped from orbit to fight on a hostile planet. The book may irritate us wishy-washy liberals with its glorification of war – but it's dismayingly readable.

'Today we're going to show you eight silent ways to kill a man.' Another military SF novel, this time highlighting the folly and horror: Joe Haldeman's *The Forever War*. You read on with guilty curiosity, wondering what the eight ways are; you don't find out, but the story has you gripped.

Are violence and tension the keys? Not necessarily. Puzzlement can be enough: 'He doesn't know which of us I am these days, but they know one truth.' What's going on here? Alfred Bester's short 'Fondly Fahrenheit', one of the most compulsive SF stories ever, dances to this crazy tune of shifting pronouns from start to end. A *tour de force* ... meaning, not an act for anyone else to follow.

'The idiot lived in a black and grey world, punctuated by the white lightning of hunger and the flicker of fear.' Really an idiot? (Yes and no.) This teasing question and the short, vivid phrases suck you into Theodore Sturgeon's idiosyncratic *More Than Human*....

'to wound the autumnal city.' (Lower-case T, not a capital.) This is the Mystification and Irritation Gambit, in Samuel R. Delany's very strange *Dhalgren*. You read on, if at all, out of interest in exotic style or to find why on earth it opens in mid-sentence. Answer: 900 pages later the closing sentence wraps around to the beginning. The butler did it but James Joyce did it first.

Straight description can carry menacing promise: 'The sky above the port was the colour of television, tuned to a dead channel.' In the cyberspace future of William Gibson's *Neuromancer*, even natural things like sky are routinely described in high-tech metaphor.

Quieter still: 'When I was quite small I would sometimes dream of a city ...' Simple and haunting, especially when it emerges that in the post-holocaust opening of John Wyndham's *The Chrysalids* there are no cities....

Excite the readers. Tantalize them. Dazzle them with imagery. An effective opening does one or more of these things and lures the sucker into your story with the promise of goodies to come. Note the word 'promise': the opening hook is a request for credit, for an investment of the reader's time and attention, and the rest of the story has to deliver. (My wildly exciting imaginary paragraph, above, makes too many promises for credible delivery.)

And at the end, of course, comes the entirely different skill of writing a good last line. How many closing lines can you remember?

'For a moment I thought I knew where I was, but when I looked back' ... yes, this ends in mid-sentence with no full stop. It looks placid, but as the end of Christopher Priest's *The Affirmation* it packs a hefty punch because the line has appeared earlier, and it was unfinished the first time too, and in a twisty way this hallucinated book has become its own sequel.

'This much we have learned. Here is the race that shall rule the sevagram.' This judgement on humanity by some passing aliens at the close of A.E. van Vogt's *The Weapon Makers* has boggled generations of sf readers for the simple, disorienting reason that the mysterious word 'sevagram' appears here for the first time in the book! As with the Bester opening, this is not a trick to imitate.

Last comes the unforgettably apocalyptic but often misquoted punchline of Arthur C. Clarke's short 'The Nine Billion Names of God'. Let's see if my failing memory can recall it: 'Overhead, without any fuss, the PCW monitors were going out....'

PCW Plus 94, July 1994

Netted Again

The meshes of Internet are closing around me ... my PCW Plus piece on computer nets (May) led to the kind of response expected from tossing a rare steak into a piranha tank. Electronic mail arrived from many PCWs (my address is still ansible@cix.co.uk). Questions were asked, and here are some answers.

• 'You recommended Britain's CIX net but didn't print the address!' CIX can be reached at *[obsolete address expunged]*, or cixadmin@cix.co.uk on e-mail. I chose it because it's cheap if used in moderation – and it's possible to access CIX from a PCW. The rival lot Demon may be a better bet for heavy use, since they charge a flat monthly fee rather than per minute (but you still pay BT charges): Demon Internet Services, 42 Hendon Lane, London, N3 1TT. The cheapest solution is to be one of the countless academics, students or businessfolk who have free net access ... lucky swine.

• 'Where can I find PCW-relevant stuff on the net?' On CIX, the PCW is mentioned regularly in the conferences 'amstrad' and 'cpm', whose names I shouldn't need to explain – also occasionally in another called (alas) 'obsolete'. Conferences are designated areas of CIX where people place messages and swap chat on particular topics; on the world Usenet bulletin board they're called newsgroups, but the principle is similar. The only Usenet newsgroup that seems even slightly relevant to PCWs is 'comp.os.cpm' – under heading COMPuters, subheading Operating Systems, and specific topic CP/M. If Usenet can offer alt.sex.bondage, why not alt.grump.alansugar?

• 'How do I get PCW communications software for CIX?' People use all sorts of things. Commplus comes from Margolis & Co, 228 Alexandra Park Road, Wood Green, London, N22 4BH, and has its own support conference on CIX to deal with queries. Qterm is public domain CP/M software; a version specially adapted for the PCW is available in the above-mentioned 'cpm' conference ... er, find a friend who uses CIX? Both Commplus and the modified Qterm come with script files – preset instruction sequences which help automate getting into CIX, sending and receiving messages, and signing off as fast as possible.

Other PCW netfolk use Kermit 4.08 (public domain) or the comms bit of Mini Office Professional. I've yet to hear of anyone happily connecting via that famous software equivalent of two cocoa tins and a string, MAIL232.COM. One negative recommendation comes from a chap struggling with PCWCOMMS ('part of PCWWORKS') who can fleetingly read incoming e-mail on the PCW screen but can't transfer it to disk files for reference, and implies that this product is poorly supported. H'mm.

Remember that you also need an interface box and a modem, and that no PCW comms software is as 'friendly' as, say, LocoScript.

- 'What is the Information Superhighway?' Seemingly it's a US government hype phrase bearing the same relation to today's Internet as 1980s 'Star Wars' defence plans had to real guns and missiles. Think of it as a catchall term for numerous vague ways in which the net could evolve....

- 'Tell me some things to do on Internet.' Connect to Coke machines in American universities, what else? No joke: many bizarre objects have been attached to Internet, and it is awesomely possible for anyone on the world net to 'finger' certain Coke machines and check their stocks – also, I believe, whether the cans are properly chilled. Last Christmas one US company put its Xmas tree on line, allowing worldwide monitoring of which sets of pretty lights were flashing at any time. The mind reels.

- 'How do I do that?' I'll spare you the Coke machines. The real ways to learn about Internet's colossal resources are to use it or to read some of the guidebooks appearing in ever-increasing numbers. One introduction, 'Zen and the Art of the Internet', comes free on the net: if you're able to download it you may not need it, which is very Zen.

Two quick 'finger' examples. First you need to reach the Internet prompt – on CIX type 'run internet', which produces a prompt like CP/M's: 'ip>'. Now you can request information on any net user, vending machine, Xmas tree or toaster on Internet, by typing 'finger' and the appropriate net address. Try:

    finger quake@gldfs.cr.usgs.gov

This returns a mass of free electronic text from the US National Earthquake Information Service computer, reporting on earthquakes all over the world. Another:

    finger nasanews@space.mit.edu

... which, as you guessed, produces voluminous news bulletins about recent NASA space activity.

- 'What have you got against using smiley faces :-) to warn that a net message is meant to be funny?' Nothing at all! The style police have had a long talk with me, and I now agree that smileys are infinitely easier than taking the trouble to write something that's actually funny. I have also learned to laugh heartily at my own jokes while digging listeners forcibly in the ribs. That way, there can be no mistake.

PCW Plus 96, September 1994

Dated Information

Whenever I write a little computer program to make life easier, cans of worms begin to pop relentlessly open and wriggle all over my working hours. This time it's the fault of a writer pal whose current novel uses dates in the 19th century. Which days were Mondays in July 1882? No trouble, I said, dimly remembering that *somewhere* I had an equation that always gives the date of Easter Sunday....

Mere hours later I found the page in T.H. O'Beirne's pop-maths book *Puzzles and Paradoxes* (Oxford University Press, 1965, another monument to my habit of never throwing books away). Oh dear. It wasn't just a handy equation. To locate Easter there were ten operations to perform, involving fourteen variables, many represented by Greek letters. Here they are, de-Greeked: I bet you can hardly wait.

(1) Divide the year number (e.g. 1994) by 100, calling the quotient B and the remainder C. The 'div' and 'mod' operators in my favourite programming language Pascal are handy here.

(2) Divide (5B+C) by 19, giving remainder A (ignore the quotient).

(3) Divide 3(B+25) by 4, giving quotient D and remainder E.

(4) Divide 8(B+11) by 25, giving quotient G.

(5) Divide 19A+D-G by 30, giving remainder H.

(6) Divide A+11H by 319, giving quotient M.

(7) Divide 60(5-E)+C by 4, giving quotient J and remainder K.

(8) Divide 2J-K-H+M by 7, giving remainder L.

(9) Divide H-M+L+110 by 30, giving quotient N and remainder Q.

(10) Divide Q+5-N by 32, giving remainder P.

Actually the book gives two sets of instructions: the above is the easy one for slower pupils (me). I shudder at doing all this on a calculator, but it's straightforward to program. And the answers are P and N: Easter Sunday will be the Pth day of the Nth month. Other weekdays in the year are then easy to calculate. It works!

Well, it works if you get the leap years right. The calculation depends on the Gregorian calendar introduced in 1582, where century years are only leap years when they divide exactly by 400: 2000 will be leap but 1900 wasn't.

I titivated the program to display or print a calendar for any selected year, and my grateful pal responded, 'What about Pepys?' Samuel Pepys recorded Sundays in his famous 1660s diary: '30 September. Lord's Day. Up betimes. Did commit great naughtiness with our serving-maid, and was mightily pleased.' Unfortunately the calendar program gave these Lord's Days as Thursdays....

This was because the Protestant countries of Europe dragged their feet over Pope Gregory XIII's calendar reform – saying they'd prefer to be 'wrong with the sun rather than right with the Pope'. Most countries gave in around 1700-1701, but two lots of anti-Catholic diehards hung grimly on for another half century: the Swedes and us. Britain changed over in 1752, when Wednesday 2 September was excitingly followed by Thursday 14 September and anguished cries of 'Give us back our eleven days!' (Not just the ignorant response of peasants who didn't understand calendar reform: landlords collected a full quarter's rent despite the shorter time.)

So the acid test of any perpetual calendar program is how it displays September 1752. Mine now smugly skips from 2 to 14 without blinking, and confirms Pepys's 17th-century record of Lord's Days. Because the Easter calculation that gives those weekdays is based on the Gregorian calendar, you also need extra corrections for leap years in the Julian calendar used in Britain before the famous eleven days. This treated century years as ordinary leap years ... so 1600 was the same in both systems but February 1700 had 28 days in the Gregorian calendar, 29 in the Julian.

Obviously this program isn't universal: all those special cases like the September 1752 changeover make it highly specific to Britain. And not all of Britain all the time, because I'm too lazy to research what differences there used to be between the Latin and Celtic churches' Easter algorithms until they sorted things out at the Synod of Whitby in AD 664. Nor can I be bothered to calculate back to the Council of Nicaea, AD 325, which defined Easter and gave rise to my favourite useless mnemonic: 'There need be no error the whiles we do recall / That Easter on the Sunday after the full moon on or following the vernal equinox doth fall.'

Cans of worms, as I said. An intended hour's work swelled into days of groping through obscure reference books and dusty almanacs, and my brain hurts. Somehow computer programming always turns out like this.

PCW Plus 98, November 1994

Centenary Blues

It's strange to look back over 100 thrill-packed editions of *8000 Plus*, as it was called from its launch issue in October 1986 to the 63rd in December 1991. Being an ancient guru with a long white beard whose column began in that first issue, I've been exploring my files....

My original brief was, roughly: 'Talk about being a writer. Talk about writing on the PCW computer. Make some jokes. And never say an unkind word about that great and good man Alan Sugar, founder of the feast.' Well, three out of four isn't too bad.

October 1986. I dutifully flew the PCW flag. 'I was slumming in an IBM PC magazine, where I found the fascinating news that Amstrad PCW systems have met with a 'lukewarm response'. To translate this you need to know the subtle linguistic codes used by IBM enthusiasts. If you buy an IBM PC, that's a ringing declaration of total commitment. If you buy something else, it's a lukewarm response.'

December 1986. The column's first prediction: 'Will Alan Sugar sandbag his existing users – as he's done before, in other ways for other computers – by abandoning 3" drives altogether? [...] Is the Pope a Scientologist?' The same month recorded a rival magazine's BASIC tutorial. 'Consider the following very simple program: **20 FRED = 30** ... When RUN, all this program does is to make the variable FRED equal to 37.' Suddenly I realized BASIC was subtler than I'd thought.

January 1987. Another safe prediction. 'I have a dark suspicion that IBM versions of LocoScript are being bolted together for those who plan to swap machines and would prefer a familiar program to a more powerful one.'

April 1987. By now I was a small PCW software company. 'We will not soon forget the man who *twice* rang us in a towering rage because our package hadn't reached his desk within two days, and ditto the replacement we apologetically rushed to him. Later, his secretary returned the extra copies of the software and we were interested to find on each a RECEIVED date-stamp showing that it had arrived the day after despatch. He had been in too much of an angrily urgent hurry to bother checking his in-tray.'

October 1987. 'Hello, we want free copies of all your software,' said the phone call. 'No,' we subtly riposted. 'But we're *Amstrad*!' The obvious reply: 'Alan Sugar can afford to pay retail price for a disk if he wants one.' The crushing retort: 'Alan Sugar didn't get where he is by paying for software, sunshine ...'

January 1988. Book publishers were still technologically backward. AUTHOR: 'I can send it to you on any standard disk, 3", 3½" or 5¼", or if you've got an electronic mail link we can ...' THRUSTING, GO-AHEAD PUBLISHER: 'Oh God, just post me a manuscript.'

April 1988. Virus scares grew, but not on our turf. 'Don't panic yet. It's the users of other computers who are learning that when you sleep with a strange disk, you sleep with all its old mates.'

October 1988. More classic truth from a lesser PCW magazine. 'One of the constant grumbles in the pages of the German computer magazines is that they have to put up with programs, manuals and adventures all written in English. It's a problem which, thank goodness, we on this side of the Channel don't have to face up to.' All too true.

March 1989. Style checkers were in the news – software that says tut-tut at long words and sentences. Thus my example of program-attested readability: 'I have been. Trying to. Improve my style and. Make it. Easier. For you punters to read. Is this any. Better?'

April 1989. Aunt Davinia's advice column from an 1889 magazine about typewriters (*Remington Plus*) offered pearls of wisdom. 'The key you seek is the very wide one which you will find lies closest to you as you operate the mechanism. Contrary to your somewhat petulant implication, we consider the manufacturers to have labelled this clearly and correctly with a picture of a space.'

September 1989. Naked pedantry showed its ugly face, with a whole page on proper punctuation. 'There's no grammatical rule against slapping exclamation marks on every sentence you think is dramatic, clever or witty! However, this is the literary equivalent of laughing loudly at your own jokes while digging violently at the listener's ribs!!!'

November 1990. 'Fifty issues of *8000 Plus*, and I've appeared in them all....' This is more or less where we came in. Alas, I couldn't be in every issue from 51 to 100, but have mercilessly repeated all the above jokes and denunciations as often as I could – with asides on Adventure, indexing, rejection slips, padding, lawsuits, chain letters, e-mail, prediction, padding, recessions, research, programming, padding and of course padding. Meanwhile, champers all round and roll on issue 200.

PCW Plus 100, January 1995

Party Time

Gosh, that immense gala party for the 100th issue of *PCW Plus* was an incredible event and for some reason I still haven't quite recovered. The sight of all the past editors snaking down Westgate Street in an impromptu conga line will not lightly be forgotten by the people of Bath – but that came towards the end of the day. The following is what the fabulous Dave's Disk Doctor Service Ltd later managed to extract from my few remaining hungover brain cells.

Champagne flowed freely in all directions. A 100th birthday cake exquisitely baked in the shape of Alan Sugar was ceremonially cut by editor Dave Green (the cake's strawberry jam filling was perhaps not in the best of taste), and thanks to the miracles of videotape the bearded founder of our favourite company delighted everyone at the celebration with his hearty recorded message of greeting: 'Buy more computers, you bastards.'

Meanwhile net-surfing production editor Rebecca Lack had put the party on-line to Internet, and the congratulations of world notables crowded her PCW display ... all the way from clinton@whitehouse.gov to HM Customs & Excise (terminator@vat.uk) trying to track down an overdue VAT return from someone called Langford. Alas, Boris Yeltsin could only send an international telegram complaining that he couldn't afford a net connection, or for that matter a PCW.

Future's publisher Simon Stansfield presented each of the editors and contributors with a lovingly hand-crafted copy of *PCW Plus* in solid gold ... and then it was time for a traditional game of Hunt the Missing Address Mark under all the tables and carpets of the party room. Other mirthful party games followed, including that old favourite 'Outguess the Spelling Checker' – who would have imagined that you could blow out a PCW's RAM chips by trying to spell-check our previous editor Mary Lojkine? By this time I was feeling extremely cheerful, and when I recounted my own spell-check favourites ('And for my name it suggested LANDLORD, ha ha ha!!!') the room fell silent in respectful acknowledgement of my famous wit.

It was wonderful to meet all the other former editors who formerly had just been names: I hadn't realized the astonishing extent of erudite Rob Ainsley's repertoire of off-colour jokes about Mallard Basic and Z80 assembly language, or that Steve Patient could perform such a hilarious impersonation of a malfunctioning PCW 9512 daisywheel printer, or indeed that Sophie Lankenau had that interesting tattoo. Nor had I expected such a splendid Charades performance from the absent Ben Taylor and Sharon Bradley, as they acted out the whole of that celebrated and much-reprinted CHECK3.BAS listing in Basic. 'Sounds like ... sounds like ... could it possibly be `IF UPPER(ASK$)="Y" THEN PRT%=1`?' How we all roared.

Howard Fisher of Locomotive Software was present amid the merry throng, relating anecdotes of how customers who'd encountered slight learning difficulties with LocoScript had amusingly mailed him scorpions, anthrax bacilli, venom-encrusted boomerangs and, on one droll occasion, a small tactical nuclear weapon.

Some more serious discussion took place as well – the technical highlight being a virtuoso demonstration of how to widen a PCW's three-inch drive to 3½" using only a broken bottle, a fishknife and three cocktail sticks. 'Don't try this at home,' warned the staff writer who performed this much-admired operation: 'It was only to make it more exciting that I worked blindfolded and with the mains power turned on.'

Next there was a moving special presentation to an ancient and white-bearded columnist who dated back to the antediluvian era of the magazine's first issue ... myself, in fact. The presentation consisted of – beautifully etched on a slab of lead crystal – the 34 paragraphs which owing to a typesetting software problem had been omitted from my piece in issue 52 (January 1991), thus giving everyone the impression that I had written it during a severe attack of delirium tremens. Tears came to my eyes as I accepted this memento from our kindly editor, and fell over.

The only shadow on the day's merry proceedings was that, as kindly Bath paramedics assisted me aboard a Great Western Shuttle express train and medically advised me not to come back, I realized that I'd somehow forgotten what had actually happened and would have to make it all up. Just like a tabloid journalist, really.

PCW Plus 102, March 1995

Mysteries in the Mail

One hazard of writing professionally is that strange people read your words, and may write back. Thus when I once wrote about military technology I was accused of being a wicked Commie, for suggesting problems with 'Star Wars' laser defences (then the flavour of the month in America). When I mocked predictions of Armageddon, a concerned believer sent me his definitive proof that Christ would return and the Universe reach its sell-by date on 26 June 1987. (I haven't heard from him since.) And some satirical comments about UFOs brought the bitter response that I was in the pay of the international C.I.A. conspiracy and receiving wads of money to help Cover Things Up. If only I were.

Strangest of all is the correspondent known as Rachel Oliver. Her career (she is probably not a 'she', but let it pass) began with idiosyncratic letters to my pal Brian Stableford, author of such fine sf novels as *The Empire Of Fear*. Presently the curse spread to another author of repute, Colin Greenland, and then to me ... and I was not the last.

The letters of Rachel Oliver (alias Siobahn [sic] Munster, Amanda Haertel, Sylvie, Susan Illegible, C. Judd, Penelope Garrard, Siobahn O'Connor, Kate McGowan, Jane Smyth, etc) usually claim to be from a pre-teenager suffering from leukaemia or another dread disease, writing from nonexistent addresses, who has a precocious grasp of scientific jargon and invariably suggests sf plot ideas. These are often traceable to articles in *New Scientist* – and famous science journalist John Gribbin reports by e-mail that he too receives letters from this versatile girl, sent care of *New Scientist*!

Brian Stableford was amused by young Rachel's cheerful inconsistency: 'Me again,' a familiar-looking letter might begin, though signed with a name she'd never used before. Colin and I felt more churlish: it seemed a slight insult, expecting us to swallow all those false names when every letter had the same handwriting, or recognizable tatty typescript. We guessed that 'Rachel' must read *Interzone* (Britain's sole professional sf magazine): we all wrote for it, and our addresses had appeared in its classified-ads columns as we tried to flog off our remainders.

During 1994, the frolic grew darker. The BBC received an ill-typed letter, drivelling about the cosmos in a familiar style – signed 'Colin Greenland', with his return address. Channel Four TV wrote to me rejecting an embarrassingly poorly presented sf proposal called *Hitler's Dimension*, supposedly by David Langford. The BBC did the same with a further submission 'from' a new player in the game: Stephen Baxter, another *Interzone* author whose address 'Rachel' didn't know ... so she'd given *Interzone*'s instead.

Was 'Rachel' a very, very stupid admirer who thought this impersonation would help our careers, or was she maliciously trying to damage reputations by sending out badly-presented rubbish under authors' names? Something had to be done.

What outsiders may not realize is that sf authors and critics mostly know each other. Information was pooled. Electronic mail hummed over the Net. Turning a blind eye to the Data Protection Act, sf convention organizers and magazine editors searched their databases for clues in the general area from which 'Rachel' sent her letters (York, Malton, Norton-in-Derwent). Brian, we began to realize, had never been impersonated because 'Rachel' knew that among the zillions of fake addresses on her letters, one of the earliest ones sent to Brian was genuine. After I'd spent a long afternoon with postcode and phone directories, we had a name, an address and even a phone number. I tried the number. It rang, but I didn't wait for an answer.

What would you do next? There were several wicked temptations, such as setting my Internet software to dial that number at 3am daily and make screeching fax noises at whoever picked up the phone. But I virtuously refrained. Colin's lawyer friend suggested, inevitably, that we should give money to lawyers – get a solicitor to write a terrifying letter. Economically, we wrote our own letters: Brian's was tactful and mine slightly stroppy. Months of silence followed. We wait nervously for developments. Did we do the right thing? Should we have called in the police? Jitter, jitter.

Yes, that was another aspect of the wonderful world of freelance writing! Don't all rush to give up the day job, now.

PCW Plus 104, May 1995

Urban Folklore

You keep hearing these infernally plausible modern myths, like the tale of the woman whose cat got into the washing machine – and so she dried it off in the microwave. Hordes of such anecdotes, sometimes called Dead Granny stories after one of the most famous ones, are in circulation. Almost always, they 'actually happened' to a FOAF (friend of a friend) of the person telling the tale.

But I'm growing bored with traditional urban myths. Country yokels already have their own new, modern beliefs: 'Crops do wither and leaves do fade, When EC intervention cheque be delayed.' I call on readers to experiment by spreading these fresh myths and superstitions, and seeing how far they travel. This may entail long, arduous hours gossiping in pubs, but think of the pride we'll enjoy if one of the following snippets gets into a magazine report on New British Folklore or yet another of Jan Howard Brunvand's books about urban mythology.

• One for hard disk users. James Thurber's mother believed that electricity leaked from empty light-sockets. Similarly, data from a PCW's hard disk can escape from floppy drive slots. (Especially when you've raised the hard disk's internal pressure by filling it nearly full.) Safeguard your novel-in-progress by taping up the disk slot or slots when not in use!

• Computer screen flicker can be reduced by magnetizing the power lead.

Repeatedly draw a strong magnet along the last foot of the wire for half an hour every Sunday morning. For best results, align your PCW facing magnetic north. (There are hi-fi fanatics who claim to get better sound quality by tying knots in the cables, so this sounds extremely plausible.)

• One of the 'bullet' characters in LocoScript 3 is useful when printed near the top right-hand corner of your tax return. Position it correctly and you will receive Windfall Bonus tax coding (all allowances multiplied by 14) when the form is processed by the Inland Revenue computers. Tax returns issued to MPs, senior civil servants and members of the House of Lords are preprinted with this mark.

• Satanic messages may be detected in the *PCW Plus* editorials by reading them backwards as a coded acrostic. In recent issues, fundamentalist researchers have uncovered the diabolical, subliminal commands 'SUBSCRIBE', 'BE A DEVIL, BUY TWO COPIES' and 'WOULDN'T A LIFE SUBSCRIPTION BE NICE?'.

• A PCW-based astrology program reports that no matter what your sun sign, when walking up the western side of Tottenham Court Road in London it is deeply unlucky not to cross to the opposite pavement before passing the Scientology shop.

• Postmen leave coded rubber bands on doorsteps as a contemporary version of traditional tramps' marks. For example, an arrangement of five rubber bands signifies: 'Householder has good hearing – put the "could not deliver, ha ha" card through the letterbox VERY QUIETLY.' As a writer, I think I've deciphered a further pattern as meaning: 'Occupant received another rejection slip, poor sad fool.'

• Despite contrary claims, there is a computer virus going the rounds of PCWs. Known as 'Titter', it infests spelling checkers and then their users, causing uncontrollable tittering at extremely unfunny discoveries: 'Titter titter titter, when I typed in "committees" the spell-checker suggested "comatose"! Laugh, I thought I'd die. And "Ansible Information" came out as "Unusable Information" ...' In extreme cases, sufferers write whole articles full of this material in our national press before kindly men in white coats administer the Prozac.

• Last year a certain hard-working author, though slightly muzzy and hungover, was trying to gobble a snack lunch, revise a book with an urgent deadline (much rapid changing of PCW disks), and deal with phone calls – all at the same time. When the rush was over he had an eerie, spectral sense that something was wrong: had he somehow put a disk in twice without taking one out? Everything seemed fine, though, and he forgot about that odd feeling until weeks later the PCW disk drive began to buzz. A fly crawled out, and then others. The repairman had to don a surgical mask before he could face removing what remained of a very elderly quarter-round of accidentally inserted corned beef sandwich squashed up at the far end of the disk drive.

Of course it's true! It happened to a friend of a friend. Or one of his friends.

PCW Plus 106, July 1995

Gadgets of Yesteryear

Isn't there a certain bizarre charm in outdated technology? I still savour my collection of obsolescent glories like old slide rules.... Remember slide rules? Or the 1950s *Astounding SF* cover showing a kerchiefed space-pirate swarming through the airlock with a slide-rule sinisterly clenched between his teeth?. Ah, nostalgia:

the big scientific rule with log-log scales, the miniature circular one, the telescopic helical model that squeezed out an extra decimal place of accuracy by wrapping a five-foot-long scale around a cylinder ...

For those wondering what proto-yuppies used to carry before cellphones and laptops, the Langford collection has the answer: the Swiss Precision Mechanical Pocket Calculator. It's a matt-black tube like an expensive camera lens, with hordes of adjustable slides and a handle on the end. You set up figures on the slides and add them by a mere turn of the handle. Subtract by turning the other way! Multiply by turning ... yes, you're ahead of me there.

There are also some nifty facilities for shifting decimal places: as Erich von Däniken might have phrased it, it's hard to believe such things were known to the primitive, cave-dwelling craftsmen of 1966.

But my collection's place of honour goes to the Ediswan High-Voltage Healing Box, vintage 1933, enabling you to commit high-frequency healing in your own home. Its front panel is a nostalgic vista of bakelite, with exciting knobs, a socketed handle on a lead, and strange glass electrodes held by clips in the velvet-lined lid.

You simply slot your favourite electrode into the handle's socket, turn on, and press it relentlessly against the Afflicted Part. I offered this opportunity to all my friends, who diplomatically dived out of windows. The lonely experimenter thus had to test the device on himself.

Switching on produces a hellish racket from an induction coil inside, an eerie violet glow in the glass electrode and a prickle of tiny sparks where this touches the aforesaid Afflicted Part. This fizzy sensation, accompanied by a paralysing reek of ozone, must have persuaded users that jolly beneficial things were happening.

The Box comes with a catalogue of tempting accessories – thirty-one specialist electrodes for all medical contingencies. My kit has only the bare essentials, alas: the puny Surface Electrode 'for use on any part of the face, body or limbs'; the appropriately-shaped Rake, 'very effective for Falling Hair, Dandruff ...'; the Metal Saturator, a chromed tube that bypasses the usual route through gas-filled glass to zap patients directly with 'a very strong current which gives powerful tonic effects'; ... and, most fearsome of all, the Fulguration Electrode.

This uses the principle of electric discharge from a sharp point to generate showers of vicious little sparks which 'deal with corns, warts and similar growths'. Having tested it very briefly and uttered loud opinions, I suspect that 'similar growths' may include fingers.

Luxury extras begin with the Roller Electrode, ideal for use when the Surface Electrode sticks and jerks in its passage over terrifiedly sweating or carbonized flesh. The Double Eye Electrode has twin cups allowing both eyeballs to be simultaneously convulsed. And with the Ediswan Ozone Inhaler, 'a mixture of pure Ozone and Pine Vapour is driven right to the back of the nose and down into the lungs'. Breathtaking!

Some specialist electrodes I'd rather not go into, or indeed vice-versa; imagine, if you will, the Nasal, Urethral, Rectal, Prostatic and Dental Cavity models plying their trade. I keep remembering an old *New Scientist* headline competition: ALTERNATIVE HEALER USED BARBED ELECTRIC ENDOSCOPE – SHOCK HORROR PROBE.

The instructions say the Box will cure everything from Abscess to Warts, including Alcohol and Drug Habits ('For Cocaine users a mild current applied to arms, legs and soles of feet, until the skin is reddened'), Brain Fag, Dropsy, Female

Troubles, Obesity and Stiff Neck. Cynics might wonder why the magic current, so good at making boils, goitre, piles and warts shrink away, has an entirely opposite effect when applied to Breast Development or Impotence.

Funny you don't seem to see this on sale any more.

And the moral, alas, is that all too many computer snobs out there regard the PCW as fit only to stand on the museum shelf next to that Healing Box. Well, they both still work, but one of them does seem a shade more useful!

PCW Plus 108, September 1995

My Fantasy Life

One of the slight shocks of getting into the literary business is realizing that 'major' reference books are produced by ordinary human beings who frequently sneak out of the back door of their ivory tower and visit the pub....

I'd written quite a few encyclopaedia entries all by myself, some of them even accurate. But before 1992 it hadn't become a habit. I could give it up any time. Until the second edition of *The Encyclopaedia of Science Fiction* (ed. John Clute and Peter Nicholls) began to move slowly and painfully towards publication, like a goat through a python. Too many of my sf friends were involved. In a fit of utter folly I pronounced the dread phrase which I recommend others to avoid: 'Hey, I'll volunteer to look through all the text on disk and see if I can spot any mistakes!'

The finished encyclopaedia ran to 1,300,000 words. Months vanished from my life.

But it was still throughly exciting, helping with the most important reference book in the sf field. (Like the slimmer first edition of 1979, the second edition won a Hugo award – the Oscar of science fiction – in 1994.) Hordes of my corrections went into the text, and the editors gave me a nice credit in the acknowledgements; even better, the publishers sent a free copy, not easily come by when the book costs forty-five quid. I felt a certain share of reflected glory.

Then came news of a companion volume, *The Fantasy Encyclopaedia* edited by John Clute and John Grant (the latter having been Technical Editor on the first book), who invited me to come aboard as a humble Contributing Editor. This was sheer madness, so naturally I said Yes at once. Even now the huge tome is being written, to a tight deadline....

Of course it would be almost impossible without computers and the net. I've been shuffling documents between PCW and IBM disks, squirting text to and fro by e-mail, and using the Internet 'telnet' facility to probe the US Library of Congress Information Service and the National Library of Scotland on-line catalogue, in hope of sorting out bibliographies. (The telnet addresses are locis.loc.gov and opac.nls.uk, if you're interested.) Luckily Usenet is full of helpful people: in the rec.arts.sf newsgroups I bumped into the net personage known only as 'Ahasuerus the Wandering Jew', who actually loves researching awkward bibliographic information. Praise be.

One of the great things about the sf volume was its 'theme' entries, little essays on standard sf devices like ANTIGRAVITY, MATTER TRANSMISSION and TIME TRAVEL. Fantasy is not so easy to define or subdivide like this. In the jargon, there's a shortage of critical vocabulary, and the editors have been heroically striving to borrow or invent names for themes which exist out there in the mists of fantasy but have never been properly classified. One slightly tongue-in-cheek example: PLOT COUPONS are frequently encountered in unimaginative fantasy novels, the sort where the plot consists of travelling up and down the map of a generic FANTASYLAND collecting the coupons (swords, rings, grails, talismans, etc), with the characters needing the full set before they can send off to the author for the ending.

The capital letters above are because all cross-references in the encyclopaedia are indicated in small capitals. After writing a few too many entries you find yourself thinking in these cross-references about mundane things like hangovers (see ACHILLES' HEEL), the bank balance (see WRONGNESS), the bank manager's office (see DARK LORD; BAD PLACE), and the awaited cheque from the publishers (see NEVER-NEVER LAND).

Meanwhile the brain gets cluttered up with the litter of becoming an instant expert on dozens and dozens of fantasy writers whose entries need to be researched and written. In spare moments I've also been trying to perfect some database software that will check all those cross-references and make sure that they all refer to theme entries that actually exist. This program scored an early success when it reported numerous cross-references to a nonexistent entry entitled SHADOW, which obviously had to be added since so many contributors felt it necessary!

And that, dear readers, is why I'm gibbering and twitching over the keyboard with all these straws in my hair. Sanity will be restored some time in 1996 (see IMPOSSIBLE DREAMS).

PCW Plus 110, November 1995

Life Forms

One of the enduring time-wasters of computing is the 'Life' game, invented by British mathematician John H. Conway in the 1960s and popularized through Martin Gardner's famous 'Mathematical Games' column in *Scientific American.*

Life is sometimes called a solitaire game, but it's a no-player game. All you do is choose the original pattern. The rules of Life then take over, causing the pattern to evolve in often surprising and beautiful ways....

Life is played on an infinite grid of squares or 'cells' – but infinity comes expensive and much can be done on a PCW screen. Cells can be either 'dead' – empty – or 'live', marked with a character such as a capital O. At each move, the rules decree which cells will be dead or alive in the coming generation.

Here are the rules. Each cell in a square grid is touched by eight neighbour cells. If fewer than two of these cells are live, the central cell dies – death from exposure. If two are live, the central cell is unchanged. If exactly three surrounding cells are live, a birth takes place: the central cell becomes live even if formerly dead. If four or more surrounding cells are live, the central cell dies of overcrowding.

These life/death calculations are made for every existing cell, and all resulting births and deaths are considered to happen simultaneously. Then you start again, to calculate the next generation of Life.... Imagine the horror of pre-computer days, when it was all done using tiddlywinks on sheets of squared paper! This is why Life is an absolute natural for computers.

Some Life patterns are static ('still life'): if four live cells touch in square formation, the resulting 'block' is stable. Some oscillate: three live cells in a row make a 'blinker', which flips forever between horizontal and vertical thanks to two deaths and two births each generation. Some patterns explode into exotic, symmetrical shapes. Some die out quickly. Some can move around:

```
 0
  0
000
```

This is the 'glider', whose shape wobbles and recreates itself at each fourth tick of the clock, shifted one cell diagonally from where it started. It can travel forever down the diagonal – but other patterns may eat it or reverse its course.

At this point, do you remember the secret life of the PCW program MAIL232.COM? Start it up, press F3, select 'Transfer as ASCII', press EXTRA and P ... and the screen becomes a blank Life grid. You can move around with the cursor keys, turn cells on and off with Return, and set the Life rules ticking away by pressing the space bar. See what happens to a row of ten live cells at the centre of the screen....

But it's awfully slow. This is why Life is a perpetual programming challenge, a wheel that keeps being re-invented to incorporate better tricks. Obviously you need to do a calculation involving eight neighbouring cells for every cell in the grid. No, wait, it can be speeded up by performing the calculations only for that part of the grid where action is happening. The rules of Life mean that births must happen next to existing live cells, so the 'active' area is the smallest rectangle that includes all current live cells plus a one-cell border of dead ones. Judging from its slowness, the MAIL232 freebie doesn't take advantage of this. Where's that BASIC (or Pascal, or assembler) disk? First, check the public domain catalogues.

Life patterns can be exotic. There is an amazing 'glider gun' that creates endless moving gliders. There is a 'breeder' pattern, too big for the PCW or indeed my PC, which generates endless glider guns. Conway has proved that by using Life patterns too vast to be run on any physical computer, you can carry out any possible computer calculation whatever (e.g. check out Fermat's last theorem). The mind boggles.

Still interested? Three good books are *Wheels, Life and other Mathematical Amusements* (1983) by Martin Gardner, which contains the best popular treatment of Life; *Winning Ways vol 2: Games in Particular* (1982), by Elwyn H. Berlekamp, John H. Conway and Richard K. Guy, which in one achingly condensed chapter takes Life from first principles to the brain-bursting limits of mathematics; and *The Recursive Universe* by William Poundstone (1985), which carries on from there. Happy headaches.

PCW Plus 112, January 1996

At Another Crossroads

Once upon a time, a tiny software company called Ansible Information saw enthusiastic reviews of the PCW 8256, and bought one to play with. Base motives were at work: after selling add-on software for earlier computers, Ansible hoped for new worlds to conquer. Could its mighty programming division – that is, the miscreant Langford – adapt to the PCW?

The good news was that LocoScript contained just what we needed: a non-printing text mark, Reverse Video, that could be used to mark up documents for our trusty indexing program AnsibleIndex. (This extracts an alphabetical index of marked words and phrases, complete with page numbers.) More good news was that our favourite program compiler – Borland's Turbo Pascal, used to write the original AnsibleIndex – came in a CP/M version.

The bad news was that LocoScript 1's document file format was insanely complicated. Weeks of forensic analysis were needed before our software could reliably read these cryptic documents and spot the page breaks. The PCW AnsibleIndex began to sell like, well, fairly tepid cakes ... and, seemingly mere days later, along came LocoScript 2 with a *different* insanely complex file format. Back to the padded drawing-board! We vowed to pull our own heads off rather than go through all that again for LocoScript 3 – at which stage, kindly Locomotive Software allowed us a peep at their own specification sheets.

Next came the horror of the 3½" disk drive. One day someone urgently wanted software on 3½" disk ... embarrassing, since we only had a 3"-drive PCW 8512 (our 8256 had died, while the 9512 fell victim to Alan Sugar's famous practical joke whereby tripping over the printer cable blows the memory chips). In a fit of insanity I wrote software to handle PCW-format 3½" disks in an IBM PC drive, and copied our stuff from the 8512 via cable link: a triumph of penny-pinching dedication over common sense.

Now we're seeing magazine previews of the fabulous new PcW16, and biting our nails. Most owners of ageing PCWs will be asking themselves: 'Is it really a PCW if it doesn't have LocoScript?' (Many may opt for a cheap second-hand IBM PC and LocoScript Pro instead.) We small software developers have different worries, as follows:

First, the PcW16 won't run CP/M programs like AnsibleIndex. Time to chuck out that Pascal development kit and redo everything from scratch in assembler – a fearful pain in any case, and doubly so for this new and inscrutable operating system devised for Amstrad by Creative Technology. Will the system's function calls be documented for developers? Who knows?

There is talk of CP/M emulator software being released one day ... but promises are cheap, and in any case CP/M programs will be hard to market for the PcW16: 'I have to buy this emulator thing *as well*?' No one ever got rich selling PCW comms software, precisely because the PCW needs an additional (non-cheap) serial interface before it'll talk to a modem.

Second, the PcW16 seems designed to discourage use of anything but the supplied programs. It has a flavour of the Amstrad NC notepad computers, where you can't change your software at all. Yes, in theory, it's possible to load new programs into the PcW16 'flash RAM', a sort of permanent M drive, and run them from there. But the flash RAM apparently has to hold the operating system, the supplied software, and all your working documents, fonts, spreadsheets and databases, with the 3½" drive being mainly for backup and transfer. People won't want to keep this precious space cluttered with extra programs ... but repeatedly installing and deleting them sounds like a chore.

Third, 1995 ended with ominous rumblings about the future of Amstrad

Consumer Electronics. Sources say the trouble centres on Amstrad's budget PCs, which in today's viciously competitive market weren't quite 'budget' enough. The PcW16 could well be unaffected ... but unless it quickly grabs a large market share, it could equally well be short-lived. In which case, investing months of software development time will have been a mistake.

And I still don't know whether the PcW16 word processor (imaginatively named WORD-PROCESSOR) uses a document format suitable for our own speciality of add-on indexing software. We stand dithering at the crossroads. As someone says in the movie *Things to Come*: 'The PcW16 ... or nothingness? Which shall it be, Passworthy? *Which shall it be?'*

PCW Plus 114, March 1996

• *When even LocoScript Software decided not to bother with the PcW16, Ansible Information breathed a sigh of enormous relief and followed suit.*

Unspeakable Researches

Secrets of writing? At a recent SF convention Christopher Priest gave a presentation called 'The Prestige Workshop', explaining the arcane research into things like nineteenth-century stage magic that lay behind his nifty novel *The Prestige*. Which reminded of my own recent researches....

I'd been invited to write a short story imitating H.P. Lovecraft's 'Cthulhu' horror stories. (Lovecraft died in 1937, but is still a cult author.) These tend to be about dread creatures from beyond space and time, so appalling that mere sight of them blasts one's mind, sort of like Alan Sugar. Dear old Lovecraft wrote about them in a special style all his own, best described as eldritch, miasmic, nameless, blasphemous, gibbous, and all too fond of the foregoing adjectives.

How to do Lovecraftian horror without just sending up his style? I looked up his story 'The Call of Cthulhu', which had always intrigued me: following a 1925 earthquake, an unpleasant city emerges from the Pacific deeps, and its architecture is eye-hurtingly inhuman. One hapless seaman chased through this labyrinth of slimy stones just vanishes into an angle of rock that somehow reverses itself like an optical illusion.

Right, I thought, I'll tell the story of how this chap came back – and opened *The Graphic Work of M.C. Escher*, because Escher actually drew the impossible geometries that Lovecraft wrote about. Imagine an Escher building rotating on a graphics screen until one of its angles reversed and the man emerged. No, said the story guidelines: the action must happen between the World Wars, so computers were out, and the relevant Escher prints weren't done until the late 50s. Think again!

What remained of my first idea was a setting: Lovecraft's imaginary Miskatonic University in Arkham, Massachusetts. The story must take place in the 20s and 30s, a time when modern physics was making exciting leaps. Aha: my degree is in physics, so the narrator would naturally be in the science faculty, giving the idea of Lovecraft's evil 'Cthulhu gods' somehow manipulating physics research to their evil ends. I began to see how the story finished, without knowing its beginning.

Rather than invent needed street names in Arkham town, I remembered an obscure 1979 volume called *An Atlas of Fantasy* by J.B. Post, which contains maps of imaginary lands like Oz and Narnia and Middle-Earth – and a street plan of Arkham, including how to get from the university to the asylum. The trick is never to throw books away ... and to catalogue them all on your PCW!

The real stroke of luck came when (after a long hunt in the Langford Book Graveyard upstairs) I traced a slightly cranky volume called *The Fourth Dimension* by C. Howard Hinton, M.A., of Princeton University. This teaches how to use coloured wooden cubes to visualize four-dimensional objects like hypercubes – or, I realized, Lovecraft's impossible angles. My edition of Hinton was published in 1912, so there would surely be a copy in Miskatonic University library in the 20s. Martin Gardner's pop-maths collection *Mathematical Carnival* (1977) gave more background on Hinton's cubes, including a warning letter from someone who'd found them dangerously obsessive and hypnotic.

So now I had a university professor who by obsessively visualizing the Hinton cubes managed to contact that man who'd vanished into a geometrical paradox. What next? Having been outside space and time (aha! the story title could be a line from Poe's verse, 'Out of Space, Out of Time'), the unfortunate chap would somehow bring back insights into the nature of the universe. Time to look up Isaac Asimov's nonfiction *Asimov's Chronology of Science & Discovery* (1989) to check on the timeline of 1930s physics research. Aha again! After that, a few dates and times needed confirmation: the handiest reference was Stephane Groueff's *Manhattan Project: The Untold Story of the Making of the Atomic Bomb* (1967).

Then I wrote the story, whose general drift you may now be able to guess, but which still had surprises for me. In fiction, research books provide scaffolding; the final product always has an unexpected shape. I surprised myself by extracting two exceedingly yucky horror scenes from the above (one partly based on memories of a Holbein picture in the National Gallery, which I have on CD-ROM), and by needing to look up Samuel Beckett's *Waiting for Godot* for advice on delirium.

There probably has to be a better way to research stories, but it works for me.

PCW Plus 116, May 1996

Critical Mass

So there you are, staring alternately at the PCW screen and a tall stack of rejection slips, and wondering about a change from sending out the stories and articles about which editors are so wilfully obtuse. How about ... writing book reviews? It seems so straightforward: read a book (if you don't read lots of books anyway, the writer's life is not for you), say what you think of it in a few wittily trenchant words, and *get paid* for this.

Of course there are complications. Most newspapers and magazines find unsolicited book reviews approximately as welcome as the Great Beast 666 at a vicarage tea-party. Unless you have some kind of literary clout (a published book or two helps, as does being a well-known DJ or serial killer), the way is hard and stony. Judging by how I got my one stint of regular reviewing in a national newspaper, the best bet is to buy lots of drinks for pals with literary connections, in case one day *The Guardian* rings and asks them to recommend a reviewer.

One friend believes in wearing down magazine reviews editors by camping in their office until they give you a review book just to make you go away – but this technique should be used with caution. It's the old Catch-22 of publishing: work must be pursued with savage persistence (if you don't ask, you don't get), but this very persistence can easily make editors so tired of you that they pronounce a technical term of literary criticism which goes, 'Sod off!'

But suppose you have at last acquired a coveted review copy. Do you rush eagerly to read it? Problem one is that the reviews editor has probably assigned you some ghastly volume you'd never normally dream of reading. Problem two is that even if it's a novel you coveted, the book is no longer just a book. It has become homework. Afterwards there will be a test, whose single essay question counts for 100% of the marks.

Right, you've read it. With any luck you feel a surge of joy and energy at the discovery that you actually have a few opinions about the thing. You switch on the PCW ... but hang on, here's this commissioning letter that came with the book. Better check the arrangements. A cold shock of horror afflicts your innards as you discover that the finely judged essay you planned to write, lovingly examining all aspects of style and content, has got to be crammed into 150 words.

It is advisable to do the cramming. Reviews editors hate writers to go over length. They also hate writers who deliver under-length copy. But they do show their appreciation of those who write exactly to length, by cutting it some more. You think it's impossible to cut your perfect, condensed prose any further? Fear not: the hellish skill of a trained editor can always find a way, usually by removing either the one phrase without which a sentence becomes meaningless, or the punchline of your favourite joke.

So – at last the work is done! A polished, gemlike review, for which you will receive a tiny but gratifying sum of money. In exchange you assign the publishers ... blimey, what's this? Full world rights in perpetuity, denying you the right to reprint the thing even in your own Collected Reviews? Time for an enjoyable argument with the reviews editor, who will point out that the publishing company needs world rights because they might want to reissue the magazine on CD-ROM or the Web anywhere in the world. Fine, you say, but why insist on exclusive rights forever? Why is it necessary to debar you from ever re-using the piece yourself? The reviews editor mutters something about having to follow Big Multinational Company Policy, as enforced by Mr Genghis Khan of the Contracts Division. Be afraid. Be very afraid.

(Toadying footnote: our very own *PCW Plus* publishers, Future Publishing, have a substantially more humane approach. Well, slightly. Well, a bit.)

At last the review is delivered, and with any luck at least 80% of it will be published, with several of the sentences appearing in the order you wrote them. Congratulations: you have become a professional reviewer. Now it is time to send in your invoice and try to get paid. This is where the fun *really* starts!

PCW Plus 118, July 1996

Baron Munchausen Remembers

'Ten years of PCW Plus!' cried the famous explorer and raconteur Baron Munchausen. 'This calls for champagne! Dear me, I remember those old days of

1986, when I made my way through the impenetrable and life-threatening Dratsma Jungle in search of the legendary Graveyard of the Missing Address Marks.'

'Did you find it?' I asked incautiously.

'A Munchausen never fails – though in that bug-infested maze, dark peril stalked me at every turning. My only weapon was a crude but serviceable copy of LocoScript 0.5. Again and again savage tigers attacked me. But as each fearsome beast was in mid-spring ... I would dextrously hit the Erase function key and blast it into Limbo!'

'Fancy that,' I said.

'Then I reached my goal. Imagine the awe and majesty of the scene, in that ancient forest clearing where all the address marks from damaged PCW disks crawl away to die. Some of these tragic relics were so old as to be in runes or cuneiform. But before I could fill my pockets with the priceless treasure – there was a tremendous roar, and I knew that some fearful beast of the jungle threatened me!'

'Something usually does, at about this point,' I murmured. The Baron gave me a hurt look, and I hastily refilled his glass.

'It was an appalling monster – oversized, overspecified, overpriced, garishly coloured and with eyes like fearsome Windows – a veritable Ponderous Critter, or PC. Death or bankruptcy stared me in the face. There was only one slim chance! Much though I regretted the cruelty to my trusty steed Joyce, I whipped out an emergency CP/M system disk and ... booted her up.'

'Er, exactly what good did that do?'

The Baron smiled. 'We Munchausens are stronger than common men. Such was the force of my boot-up that my PCW steed and I at once went flying through the air, high above the forest – clean out of danger! It was the work of a moment to ask directions from a friendly duck that came flapping by –'

'Excuse me, Baron,' I interrupted. 'Of course I believe every word you tell me, and would never so much as dream of questioning your veracity, but exactly where did you learn the language of ducks?'

'Every PCW, my friend, comes with a manual of Basic Mallard. Of course, even when I had properly set my course for home (using the duck's very clear GOTO statements), I was still in the throes of mortal peril! Falling to earth from such a height would inevitably be a fatal system crash. It was an even trickier predicament than when, as I told you last time we met, my shirt-cuff became caught in the PCW disk drive while LocoLink was running, so that I myself was horribly sucked into the hardware and converted to IBM format. How I escaped –'

I raised a warning finger. 'No digressions, now.'

'Oh, very well. Where was I? Yes, soaring high like a cannonball above the Earth! Luckily I am a quick thinker, and still more luckily a spider from that jungle had nested in my famous tricorn hat. With the aid of its silken thread and my trusty RS-232 interface, I connected to the World Wide Web and purchased numerous copies of AnsibleIndex, LocoSpell, Masterfile 8000, Microdesign, Mini Office Professional, Protext, and many more. As I had requested, these items were heaped into a large pile, on which I landed as softly as a feather –'

'– because it was a pile of software,' I groaned, looking feebly around me for a sick-bag.

The Baron frowned. 'Have I told you this adventure before? ... So all was well, except that my PCW keyboard was damaged as we landed – one keytop was battered to a PASTE. It only remained to call Amstrad Technical Support: they answered the phone at once, eager to do everything in their power for any customer, and Alan Sugar himself assured me that it didn't matter at all that I'd violated the warranty by opening up the machine to install the extra RAM and some ewes. A replacement part arrived free of charge by special courier on the very next day ...'

'Hang on there! Although I trust you implicitly, Baron, I don't think I can quite believe *that* one.'

For the first time since I'd known him, Baron Munchausen blushed.

PCW Plus 120, September 1996

• *Non-PCW-using readers may need to be told that the Amstrad PCW keyboard has special dedicated keys like CUT and PASTE; that 'missing address mark' meant a damaged floppy disk; that most PCWs came with a programming language called Mallard BASIC; and that Amstrad technical support was and is ... legendary. Yes, legendary would be the word.*

The Customer is Always Right

Another instant urban legend of the computer world was circulating recently, popping up all over Internet and even making it into occasional newspapers. This is the one about the customer who phoned the technical support hotline to complain that the cup-holder on the front of his IBM PC was broken, and please could it be replaced under guarantee? 'Cup-holder???' thought the baffled expert, and asked a few cautiously probing questions ... whereupon he was removed in strong hysterics after realizing that the relevant feature of the computer was in fact the compact-disc tray that whirs in and out at the touch of a button. This, curiously enough, is designed to take featherweight CD-ROM discs and not huge dripping mugs of coffee.

No – surely this story has to be a myth? I think so. I think we can also disbelieve the older tale of the typist who meticulously blotted out errors on the PCW LocoScript edit screen using Tipp-Ex and a little brush.

But on the other hand, I do in fact know someone who genuinely experienced another legend from the early years of desktop computers. 'You must make daily backup copies of your disks,' the office secretaries were told. Came the day of the inevitable disaster, and our chap went in to restore all the lost work from these backup copies, and was awestruck to be presented with a thick ring-binder full of carefully-made photocopies of floppy disks.

Many witnesses also attest to the existence of computer users who found an ingenious way to store important disks so they'd always be conveniently to hand. You simply fix them to the front or side of your filing cabinet with powerful fridge magnets. Now why didn't I think of that? (Kids! Don't try this experiment at home!)

Not so mythically resonant, but just as frustrating, was trying to sort out the software customer who simply could not get the hang of the Alt and Extra keys. No matter how many million times you explained the need to 'hold down Alt and press C' (to get the Copyright sign), this person's internal brainware would always rewrite the instruction as 'hold down Alt for a bit, release it because that's what the silly man obviously means, and then press C' ... shortly followed by a wail of 'I did just what you said and it doesn't work!' This is an amusing way to delude the expert at the other end of the phone line that your keyboard must be broken.

Then there was the phone enquiry about a PCW that didn't start at all. Not a flicker from the screen, not a glint from the disk-drive lights. Expert, scratching head: 'Er, are you sure it's plugged in?' Annoyed customer: *'Of course* it is! What do you think I am, stupid?' This, explained the technical-support man with twenty years' experience, was the wrong approach; and he demonstrated how the conversation should go – as follows. Expert, with great confidence: 'There can sometimes be trouble with some defective plugs. Can you look at the pin side of the mains plug to make sure it has the BS 1363A mark?' Customer: 'Hang on, I'll just go and look.' (Short pause.) 'Oh gosh, you'll never believe this, but it wasn't ...'

From the same bottomless pool of legend, there's the proud mother who chuckled fondly as her ingenious toddler fed Drive B with Rich Tea biscuits (and the slack-jawed repair man gasped, 'Crumbs!') ... the absent-minded fellow who used his document disk as a coaster, which was fine until he actually spilt some sticky, sugary coffee ... the thrifty person who re-inked a ribbon with best fountain-pen Quink, the result being like something out of a horror story by H.P. Lovecraft, all awash with blasphemous ichor ... the clot who complained of a serious, intermittent system fault in a PCW whose power came from the top of a stack of mains adaptors, three high and wobblier than New Labour policy ... the inspired genius who literally cut through all the difficulties of 3½" to 3" disk conversion, with a single stroke of the office guillotine ...

Such are the stories that the old hands of computing swap in the pub. Of course, if they heard the ones that customers tell about *them* – appalling tales of non-delivery, missing parts, failure to turn up, condescending rudeness, bug-ridden software and incomprehensible manuals written in post-structuralist jargon – why, they'd be shocked. In fact, they'd probably sue.

PCW Plus 122, November 1996

Thog's Masterclass

An early and much-appreciated 1996 Christmas present was my copy of Sarah LeFanu's *Writing Fantasy Fiction* (A&C Black, £8.99). Besides containing much good advice, this quotes me, and instructs aspiring writers to read my SF newsletter, and – best of all – recommends study of the awful warnings in the newsletter's 'Thog's Masterclass' department.

What is Thog's Masterclass? Good question. This is named for Thog the Mighty, a dimwitted barbarian hero invented by fantasy novelist John Grant ... who like me does much reviewing and copyediting, and likes to collect specimens of barbarous prose which somehow get into print. This isn't just nitpicking. Thog's focus on how things can go wrong invites a closer look at one's own sentences. It's also encouraging to see that published – even best-selling – authors commit such bloopers.

One problem is failing to visualize what's being described. 'He lifted her tee-shirt over her head. Her silk panties followed.' (Peter F. Hamilton, *Mindstar Rising*.) Or: 'The green fur made it look like a Terran gorilla more than anything.' (Michael

Kring, *The Space Mavericks*.) Or: 'Sweat broke out on his brow as he wrestled with his brain ...' (Julian Flood, 'Control'.)

It's also unwise to forget what you wrote in the previous sentence. 'Susan awoke to an absolute silence: the traffic outside the hotel had been utterly stilled. John was in the bathroom – she could hear the shower running.' (Robert Charles Wilson, *The Divide*.) Or, indeed, in the *same* sentence: 'So instantaneous and final were these lethal rays that the destructive act was over in but a few minutes.' (Nal Rafcam, *The Troglodytes*.) 'The brassy September blue overhead had been obscured by invisible storm clouds.' (Emil Petaja, *The Nets of Space*.)

The ambiguities of English make it all too easy to write sentences that can be read in a way you didn't intend – like the famous newspaper headline POLICE FOUND SAFE UNDER BED. 'She knew how to embroider and milk a cow.' (Connie Willis, *Doomsday Book*.) 'He swept the antechamber with the eyes of a trapped animal.' (Poul Anderson, 'Among Thieves') 'Something jumped in the back of Morgon's throat. It was huge, broad as a farmhorse, with a deer's delicate, triangular face.' (Patricia McKillip, *The Riddle-Master of Hed*.)

Even famous authors do it! Unlikely Geography Dept: 'She wore large bronze earrings made in an obscure country which rattled when she laughed.' (Brian Aldiss, *Remembrance Day*.) Implausible Physiological Tricks: 'His mouth, for a moment, ran liquid and then it slid, almost of its own accord, down his throat.' (Isaac Asimov, *Prelude to Foundation*.)

Want to write evocatively, poetically? Beware of excess: 'When they finished eating, they would lie silently under the blankets until sleep shuffled over the roofs to the leaded skylight and threw itself down on them, sprawling like a wanton over their faces.' (Felicity Savage, *Humility Garden*.) 'Arias plunged his blue-grey regard into hers.' (Anne Gay, *To Bathe in Lightning*.) –- that is, he stared into her eyes. 'Her very existence made his forebrain swell until it threatened to leak out his sinuses.' (Nancy A. Collins, *Sunglasses After Dark*.)

Similes are fraught with pitfalls. This SF description of a space elevator seems, er, understated: 'Just to the south of them, the new Socket was like a titanic concrete bunker, the new elevator cable rising out of it like an elevator cable ...' (Kim Stanley Robinson, *Green Mars*.) Next, horror combines with fruit salad: 'His head suddenly began to peel, the flesh tearing away from the bone in ragged strips, like a pink banana.' (Robert Holdstock, *The Stalking*) And this one makes you wonder about the author's kids: 'They were featureless and telic, like lambent gangrene. They looked horribly like children.' (Stephen R. Donaldson, *White Gold Wielder*.)

Occasionally, the problem is just thickwittedness: 'He absorbed Latin in two hours yesterday! It took me a whole year just to learn the Latin alphabet.' (Screenplay, *The Lawnmower Man*.)

In fact, authors are human and make mistakes. But Thog's bloopers are multiple mistakes, since in an ideal world the editor, copyeditor or proofreader should catch these things. Sometimes they do, only to be overruled. Consider this sentence from the proofs of Robert Jordan's doorstop fantasy blockbuster *The Fires of Heaven*: 'Elayne wished the woman would just revert to herself instead of bludgeoning her with a lady's maid from the Pit of Doom.' Alerted by the daft image of bludgeoning someone with a maid, one reader issued a warning query. And lo! the author realized that the sentence could indeed be improved, and he carefully altered 'the Pit of Doom' to 'the Blight'. Oh dear.

PCW Plus 124, Xmas 1996

• *And at this point, with mere hours' warning even for the hapless editor, Future axed the magazine. The miserable sods never even sent me a complimentary copy of the above issue with my last appearance. But a year or so later came an invitation from PCW Today ...*

Looking Backwards

There's one big question that gets asked wherever two or three PCW owners are gathered together: 'Where were *you* when you heard that *PCW Plus* had been axed?'

The answer for me is, well, the same place where I learn most news: sitting at a keyboard watching e-mail and Usenet messages come chugging down the line. The date was 2 November 1996, the place was the comp.sys.amstrad.8bit newsgroup, and I still remember reading with slack-jawed horror how Future Publishing had deployed fear, surprise and ruthless efficiency by rushing the bad news to the *PCW Plus* editorial staff just a week before the final issue 124 went to press.

At that time I felt a bit like the oldest inhabitant, wearing a long white beard and mumbling: 'Ar, what do these new users with their pansy 3½" drives know? In my day we had to mortgage our wives just for the down payment on a pack of CF2s, and *still* hunt all over the floor for the missing address marks. And when you told LocoScript 0.5 to move to the end of a 4k document, you had time to make a pot of tea, drive to the airport, go on holiday to Majorca, and come home before it finished scrolling. Tell that to today's youngsters and they won't believe you. Harrumph!'

Well, something quite a lot like that. There were certainly deep, poignant emotions associated with having had columns in the very first *8000 Plus* dated October 1986, and (despite gaps when Future decided they couldn't afford luxuries like columnists) the very last *PCW Plus* in December 1996. Perhaps it's best summed up in the simple, tragic words that later came spontaneously to my lips: 'They're never going to send me a complimentary copy of #124 now, the rotten sods.'

Really, it was a strange time for the magazine to fold, since LocoScript Software ('Ar, today's youngsters don't know the thrill of the glory days when they were Locomotive.') had just caused a splash of excitement in the PCW world by releasing LocoScript Version 4. That should have been worth a few more issues of hot debate on exciting new word processing techniques, and indeed new bugs. Experienced LocoScript-watchers got the impression of wall-to-wall panic in Dorking until the original Loco 4 could be replaced a little later with the very much more wonderful Loco 4 Release 2.

Meanwhile, with *PCW Plus* gone I couldn't whinge in public about the problems caused to my own Ansible Information by the unexpected – by me, anyway – appearance of LocoScript 4. The trouble was that our internationally unknown AnsibleIndex software reads Loco files, but was unable to recognize or understand Loco 4 documents thanks to slight but significant changes of structure. So Ansible customers who upgraded to Loco 4 were soon sending ominous letters and making doom-laden phone calls to say how we could shortly expect a visit from the Men With Big Sticks.

It was a tense time. Fearlessly I shouldered the responsibility by answering these anguished complaints and queries in suitably shifty, unconvincing tones: 'Er um, it's all LocoScript's fault!' Far away in Hastings, my business partner Chris Priest backed me up nobly by answering enquiries with: 'Don't blame me, guv, blame Langford.'

Meanwhile, the old PCW was glowing red-hot as Borland Turbo Pascal 3.0 – my favourite CP/M programming language – stalked the M: drive once again after years of slumber. Good old Howard Fisher of LocoScript had sent along a buckshee copy of Loco 4, so I could work out some of the obvious changes by analysing the documents which this produced. Unfortunately, one of the not-so-obvious changes was subtle enough to be very, very hard to identify by peering suspiciously into document files with CP/M's SID.COM and my own home-made software toolkit.

Finally, at the eleventh hour, just as the entire Ansible Customer Support group had surrounded my house wearing pointy white hoods with sinister eyeholes, and were igniting a huge blazing replica of a 3" disk on the front lawn ... the hero programming team at Loco completed their magic software developer's bible, *The Structure of LocoScript 4 Documents*. Howard instantly had a copy rushed by helicopter to Reading and parachute-dropped to me (along with several copies of a terrifying non-disclosure agreement, so please don't ask), and thus the world was saved. That's what it was like in the grand old days of February 1997. Tell that to today's youngsters and they won't believe you.

What a great man Howard Fisher is. What a pity I never had the chance to write up this episode in the hallowed pages of *PCW Plus*.

According to my own files for this magazine's fateful ten years, I did somehow manage to publish 88 columns there. Only one passage was censored by steely-eyed editors: a mini-rant about the bloody awful index in the manual for the Amstrad PPC portable, vetoed because the PPC wasn't a PCW. Early on, under a different editor, I'd got away with an entire column on the PPC. So it goes.

More in our next ...

PCW Today 8, Winter 1997-1998

Change and Decay

On the memorable day when the PCW 9512 upstairs in the spare room went bang (I think as a result of the famous Alan Sugar Surprise, whereby tripping over the printer cable wittily blows the RAM chips), my business partner Chris Priest soon found the appropriate words for the occasion.

'What you must remember, young David,' he faxed, 'is that owning a computer is very like *having a mistress*. If you keep her on the shelf and neglect her for months or years while you go flirting with younger models, she just won't want to perform for you any more ...'

It was true that the 9512 didn't see much use – I kept this one as a backup but preferred the 8512, on the well-known *Animal Farm* principle of Two Drives Good,

One Drive Bad. It's also sadly true that some computers are more equal than others, and that nowadays even the Joyce 8512 gets her bit of slap and tickle only about once a fortnight. So, like a wronged mistress, she too is taking revenge.

This began with a certain reluctance to get started on cold mornings. (Er, can we drop the double-entendres now?) My 8512 is probably long overdue for a new belt in Drive A, but still boots up happily once the room is warm enough. I like to think that this is the same problem identified by the great Richard P. Feynman in the inquiry into the *Challenger* shuttle disaster. The shuttle's O-ring seals lost their springiness in cold weather, and so to some extent does an ageing PCW drive belt.

More insidiously, in March 1998 a number of Ansible Information working disks – both Drive A and Drive B format – came down with mild lurgi. Some intermittently wouldn't boot, even though other disks did so happily; some wouldn't yield up certain files. The link seemed to be that they were all very old disks ... as though the whole lot had reached an invisible sell-by date. But they weren't quite dead, and a simple fix was to make a whole-disk copy on to a newer disk, using DISCKIT or LocoScript. Strange but true.

All this caused me to load LocoScript, which for various personal reasons I don't really get on with. One of the most boringly obscure facts of *PCW Plus* history is that in the days when they preferred contributors to fly the flag by loyally writing on a PCW for transfer to Future's Macintoshes, I used a truly grotty old word processor called SuperWriter. This had been badly converted for the PCW from a generic CP/M version and lacked 80% of LocoScript's facilities – but did two things which from a professional writer's viewpoint struck me as more important than the exotic fonts, columns, colours and pictures later added to LocoScript.

Item one was a Find/Exchange function that worked on print controls, such as italics markers. Item two: a simple macro facility allowing the automation of keystroke sequences. Using both together, you could create a little macro that with a single command changed all a document's italics or underline marks to some agreed character (strokes, asterisks, underscores) and then converted the result to an ASCII file.

Why? Publishers, scenting the opportunity to save the cost of having material retyped, were starting to ask for text on disk. Later, they wanted it e-mailed. With the exception of *PCW Plus*, my publishers asked for plain ASCII format. But in plain ASCII generated by LocoScript, all the italics and underline marks (and any others, but those two are the most important print effects in ordinary text) were automatically discarded. When I italicize a word, I want it to *stay* in italics.

There are other reasons why I don't personally fancy LocoScript, but these are probably all my fault. After all, LocoScript was designed to be the only program PCW users would ever run, on the only computer they owned. Rather than bother with people's expectations of how word processors should work, the programmers could definitively say '*This* is how things will be done.'

Unfortunately I've been sleeping around too much (oops, forgot we'd abandoned that metaphor) and have several vague expectations based on other software. In LocoScript directory lists, RETURN or ENTER seem intuitively obvious ways to select a file, but one beeps and does nothing while the other, more insidiously, does nothing and keeps quiet about it. When I get into the wrong LocoScript menu I *still* instinctively hit EXIT or STOP as logical ways out, and curse those unforgiving beeps for a bit before remembering about the unhandily placed

CAN key. But, as a special dose of user-hostility ... when you accidentally hit the far too handily placed PTR and the whole keyboard becomes a minefield of beeps, it is necessary to note that CAN doesn't work (though it doesn't beep either) and, just for once, EXIT does.

I'm sure there are excellent reasons for all this. Perhaps, when every byte counts, allowing alternative keystrokes – like pressing EXIT to, as it were, exit from a menu – was just too demanding in terms of program size. Perhaps I'm just a picky sod. (*Muffled Voice From General Direction Of Dorking:* 'He's talking sense at last!') Perhaps I'll go and lie down until next issue....

PCW Today 9, May 1998

• *It should be easily deducible that* PCW Today *was edited from Dorking.*

Take Another Look

One of my hobbies is running a little science fiction newsletter, called *Ansible,* and in a recent issue ...

(Oh, all right, let me explain. The infamous PCW software house Ansible Information was called after the newsletter, simply because I already had a bank account in the name of Ansible. The actual word had been coined by Ursula Le Guin in the 60s and used in several of her SF novels for an instantaneous interstellar communication device ... which seemed to make it a good SF newsletter title. Let me brag a moment: *Ansible* is the only British publication ever to top the Best Fanzine category of science fiction's US-dominated Hugo award, a category which *Ansible* has won three times. On the other hand, Christopher Priest points out that the title is an obvious anagram of 'lesbian' – perhaps Ursula's little joke.)

As I was saying before I so rudely interrupted, a recent *Ansible* carried a news snippet about the cutting-edge SF author Neal Stephenson, author of novels like *The Diamond Age* which explore the radical effects on society of new computer developments, virtual reality and nanotechnology. The interesting point was that Stephenson had just gone public with an announcement that he'd given up writing with a word processor, and invested in a fountain pen. Gorblimey.

Why? Because 'his laptop computer crashed, erasing a large chunk of writing.' That is, because he was too idle to make backups. Perhaps realizing that this was not the most sensible of reasons, Stephenson went on to add that the computer *made things too easy* and 'caused me to spill stuff out as fast as I could type.... When I went back and read it later, I found that I was using hackneyed phrases and sometimes writing in kind of a thoughtless way. With the pen, I tend to go a little slower and think a little harder.' (*New York Times*, 9 April 1998)

You don't have to be a crazed Luddite to sympathize a little with the problem – even if the Stephenson solution seems weirdly extreme. The lazy tendency to lose focus on what you're currently writing, let alone what you've already written, is encouraged by computers because even the best monitor display is a little bit harder to stare at with prolonged attention than mere paper. Nor is it easy to compare passages on different pages.

Another, subtler problem with word processing in general is identified by science fiction's wittiest critic John Clute in his book *Look at the Evidence* (1995).

Dissecting a novel whose title doesn't matter, Clute reckoned that it ...

> ... reads as though it had been written – as most books are today, just as this review is being written – on a computer; and if it doesn't exactly overstay its welcome the way books used to when they went on too long, it does, all the same, give off a sense that too many luxurious repetitions of the moody bits were patched into the text, *just to make sure*. (In the old days, when they were written consecutively, books *grew* too long at their top end, like buddleia; nowadays, when they can be assembled from tesseract blocks like vast mosaics, a book is likely to become too long at *any point*; and then get short again, maybe.)

Even moving blocks of text around for the best effect – so wonderfully easy with word-processing – has its hazards. You might almost unconsciously be using more urgent prose rhythms, shorter and punchier words and sentences, in the action-packed Chapter 6. Then you decide that a chunk of philosophical exposition from the more leisurely Chapter 8 needs to be moved back to Chapter 6. There's nothing wrong with the words from Chapter 8, but in Chapter 6 they subtly don't fit....

Again, we're not talking about some hideous, inherent flaw in using a word processor, but about human nature and the attractions of the line of least resistance. LocoScript and ProText don't themselves prevent you from reading your whole text closely in the right order, looking for jarring changes of rhythm, for Stephenson's hackneyed phrasing, and for Clute's quagmires of tedium where the book suddenly and locally becomes too long. Word processors just make it easier to be lazy and let your eye slide over the dodgy bits. As I wrote once in *PCW Plus*, smooth and glowing text on screen – unlike tatty sheets of typescript – can soothe your critical senses merely by looking so polished, perfect and 'finished'.

Thus many writers force themselves to take a hard look at their text by printing it out, and determinedly reading and scribbling all over the resulting pages. Terry Pratchett even has a pool of what he calls 'beta test readers', trusted fans who read the electronic drafts of Discworld books and tell him which parts don't feel quite right.

One trick to help you see your writing with fresh eyes is to change the word processor margin settings. Familiar paragraphs that looked vaguely OK as a whole suddenly appear with a new shape, revealing internal awkwardnesses of phrasing. Since I mostly write magazine columns and send them in by e-mail, I always reread the text as it appears in the completely different font, margins and background of my e-mail software – and am frequently boggled by dreadful phrases suddenly highlighted by the change in perspective.

Knot to mention sum miss takes that parsed the spieling chequer....

PCW Today 10, ??? 1998

Money, Money, Money

One of the joys of being a humble freelance writer is that you escape certain kinds of financial worry. For example, you never fret or get stressed about losing huge sums of money, because you haven't got any.

But of course I do worry about small sums, and few are smaller than the royalties that come in from obscure American editions of my work. For example, there's a little chapbook of Langford sf/horror stories published by Necronomicon Press in Rhode Island, which every so often generates a royalty cheque for something like $2.48. Once you're an established writer you always have this to fall back on – the comforting knowledge that, rain or shine, your regular $2.48 will keep on rolling in every half-year. Probably.

Of course the snag is that when you expose such a small dollar cheque (or check) to the vampire fangs of a British bank, the swingeing conversion fee which they charge on top of their cruel exchange rate means that it's literally not worth cashing the thing.

A ray of hope came when one of the Sunday newspapers recommended a banking scheme that let you accumulate dollars in a dollar account on Guernsey, with free sterling conversion. Only it turned out to be a fantastically opulent Rothschild offshore investment deal, requiring certified copies of your passport and other terrifying credentials. And the brochure was plastered with warnings that if by any chance you turned out to be a scumbag freelance who just wanted to convert little dollar cheques rather than invest the money forever, you would be hunted down by trained bloodhounds.

I finally solved the problem while visiting Minneapolis as the guest of a science fiction convention (writing this stuff does offer occasional perks). All you need do, once in the USA, is open a dollar bank account. Admittedly the cashier spent the first quarter-hour explaining that this was impossible because my British social security number was the wrong format for their accounts database – but it turned out that there was a way. I am now the proud owner of a Norwest Banks free current account, which accepts deposits by mail and comes with an ATM card allowing me to suck out the money in sterling from British cash machines. Meanwhile, the friend who was driving me around Minneapolis found a bank brochure offering rewards for introducing new customers: so she claimed $10 for introducing me, and I got the same for being introduced. Nice place, America.

Somehow, though, the bank must have deduced that I was a science fiction writer. Will I ever work up the courage to write cheques, or checks, which are embarrassingly decorated with coloured poster art from *Star Wars*? Anyway, opening a US account is now my recommended approach for writers and others who get occasional small payments in dollars. I can even flog Ansible Information software to the Americans again.

The computer enters the picture by running a spreadsheet like Supercalc or Cracker to keep track of the dollar funds over which I now gloat daily. Sometimes I think I'd have a tougher time getting by without some kind of spreadsheet than I would without a word processor. When recently stuck in hospital for a day with one arm strung up in a sort of left-handed Nazi salute (owing to a nasty hole in a wrist artery, inflicted by a milk bottle), I amazed myself by managing to write 17 pages of the current urgent-deadline work with my good hand. But the thought of doing the VAT accounts without spreadsheet software is too horrific to contemplate.

Spreadsheets were heavily involved in a rather sad task that occupied most of my spare time last year. My father, a retired accountant, had lost his powers of concentration through Alzheimer's disease and become unable to complete his own tax accounts for 1990 and 1991, on which he'd been stalled for years. The tax office went through their traditional ritual of sending increasingly inflated assessments to frighten him into completing some tax returns, and – being both unable to cope with the figures and unwilling to admit this to anyone – he delayed each payment as long as he could and then paid up. I don't like to think how much the Inland Revenue must rake in from bewildered old people in similar circumstances.

So all that year I made regular hit-and-run raids on my parents' home in South Wales, shoving financial information (from bank statements, bills, tax vouchers, every bit of paper I could find) into vast spreadsheets on the cheapest available laptop computer. By the time my father had gone into further decline, entering hospital and then the nursing home where he's now at least reasonably comfortable, I was ready with something like 40 spreadsheets representing eight years' financial data which translated into eight tax returns. The end result was that all the hideous demands for more tax were withdrawn, and several thousand pounds refunded to my father and mother instead. I still can't imagine doing all this without my own computer.

Apologies for the somewhat gloomy tone of this instalment. The outside world has been rather overshadowing the dilettante joys of freelance life and PCW punditry, making it difficult to think up exciting new Alan Sugar libels or scandalous rumours about your editor. The usual abnormal service will be resumed.

PCW Today 11, November 1998

Under the Bonnet

Sooner or later there comes the time of the Big Decision to take the bull between one's teeth, the time when a man has got to do what a man has got to do. Yes, I finally got around to changing that dodgy belt in Drive A of the ancestral PCW 8512.

A small plug here to Brian Watson of the Independent Eight Bit Association – ieba@spheroid.demon.co.uk – who in mid-1998 offered spare belts for one pound apiece, post free. I bought a couple, put them reverently on top of the PCW, and guiltily avoided their gaze for several months. As noted in *PCW Today* issue 9, the computer would still start if switched on and given a few hours to get warm enough for its withered old drive belt to regain a semblance of youthful springiness. But as the months went by, that warm-up time grew to many hours.

After this display of laziness and procrastination, it will come as no surprise that I couldn't face searching through the dusty boxes housing my run of *PCW Plus* (still incomplete – the final Christmas issue with my last column is missing: hint, hint) for one of those legendary how-to articles with witty titles like 'Belt Up!' As a qualified physicist, I should surely be able to tackle this with effortless flair, panache and screwdrivers.

Unfortunately the famous Amstrad cost-cutting construction does make the whole process more tortuous than expected, as you grope among tightly packed components. I started obsessively counting the stages and the growing piles of screws:

a) Remove the PCW keyboard connector and, in my case, the serial interface box (two screws).

b) Remove monitor case (four screws, two cunningly hard to reach). Be amazed by the great wads of dust, fluff and mummified insects which have accumulated within. Note that these screws are not all the same length ... er, which hole did which one just come out of?

c) Remove monitor stand (two screws plus two awkward lugs) ... perhaps not strictly necessary, but it makes the thing easier to manhandle. Discover how the lugs and the two halves of the case are arranged to make this incredibly difficult if, like certain Langfords of this parish, you first attempt it before step (b).

d) Remove the main printed circuit board from its slots – no screws but much paranoia at the belated recollection of all those anti-static precautions which one has just failed to take.

e) Prise off incredibly fiddly disk drive ribbon and power cables: more paranoia as the latter, in particular, seems readier to break off altogether than separate in the orthodox way. Pat oneself on the back for having correctly done this to Drive A and not Drive B.

f) Remove the Drive A mounting screws (four of these, two almost inaccessible and requiring extensive search for a screwdriver with a longer blade). Hurray, we've got the wretched drive out of the machine at last!

g) Remove the drive case screws (four) and slide out the drive itself. Blow away further startling quantities of fluff.

h) Remove the drive's front panel screws (two). It's around now that one has to go and hunt for jeweller's screwdrivers as the standard tools prove to be too big.

i) Remove likely-looking screws holding little circuit board to one side of drive (two). No use. Put them back again. Remember to tell readers to scan this saga all the way to the end before committing copycat crimes on their own PCWs.

j) Try again with the larger circuit board that covers the whole of the *other* side of the drive mechanism (two screws). It is at last possible to see where the belt actually is!

k) Remove the still more incredibly fragile connector which prevents this larger circuit board from being tilted out. Maximum paranoia as it seems about to disintegrate in a fit of petulance.... Now the circuit board can be tilted just far enough to allow access to the belt pulleys.

l) Wrench out grotty old belt with forceps. Try stretching it. Be amazed that this limp, enfeebled thing could work the drive even after hours of warm-up. Sternly suppress the Viagra jokes which spring to mind.

m) Thread the new one, the one step which proves surprisingly easy: the big pulley is nearest to hand, and once the belt is loosely looped around this, you can shove it with the blunt end of a pencil to stretch it over the smaller, less accessible pulley. Getting it twisted must be the thing to avoid ... I was lucky first time.

n) Reassemble by undoing steps (k) to (a) in reverse order, while wisely avoiding (i). It is around now that you realize what a stunningly good idea it would have been to lay the screws on a sheet of paper with detailed annotations about which goes where. The hardest thing to re-insert is the inmost drive-mount screw at step (f): getting it aligned with its inconveniently placed hole is reminiscent of pinning the tail on the donkey, blindfold.

o) Blimey, the thing actually boots up! Even from those start of day disks which I'd filed as possible cases for Dave's Disk Doctor Service, since they seemed hopeless even after the prescribed many hours of warm-up. Well, well.

p) Pausing for a small prayer that the less-used Drive B won't go the same way within the week, this would seem a good time for a stiff drink. The most appropriate cocktail is a Screwdriver.

PCW Today 12, February 1999

Pig in a Poke

You always think you can give it up, take the cold-turkey cure until at last the craving goes away, but again and again there's a dread tendency to backslide. Just one more tiny little indulgence won't do any harm....

I'm talking about programming, of course. A friend once warned me in menacing tones: 'Software is a *disease*. Never get into software.' He spoke from experience, having written his own word processor from scratch in BASIC. Although I laughed ('They laughed when I sat at the keyboard, but then I began to hack the operating system!'), it was already too late. I'd invested in Borland Turbo Pascal 3.0, which had begun to affect the punctuation of my very thoughts. It is a bad sign when you end sentences with semicolons.

Turbo Pascal was a revelation. It created .COM program files quickly and easily, and came in different flavours for all the computers I was using in the mid- to late 1980s: Apricot, PC and even a CP/M version for the PCW. Our infamous indexing program AnsibleIndex used all three, starting as an add-on for Apricot SuperWriter, mutating dramatically to work with PCW LocoScript, and eventually following LocoScript Pro to the PC.

I thought I'd finished with PCW programming after version 4.50 of AnsibleIndex, the one that handled the subtly changed document format of LocoScript 4. However, the temptation keeps returning, and any little thing can set it off.

For example, a copy of the British Amstrad PCW Club's magazine *The Disc Drive* came my way this year, and I noticed a piece by one Rod Shinkfield on something doubtless known to our more learned readers, but which I'd never heard about: CP/M's secret Limbo group, where the deleted files go.

When writing AnsibleIndex I'd naturally had to tackle access to disk groups 0 to 7. CP/M nominally has 16, numbered 0 to 15; Loco uses the second eight as its own Limbo groups. Turbo offers a neat facility for including 'in-line assembler' – machine code instructions – so you can write a machine-coded BDOS call to select any normal group through CP/M's 'Set User' function.

The occult CP/M Limbo, weirdly, is group 249 (F9 in hexadecimal). You can get to any of groups 0 to 15 with the CP/M command USER, but both USER 249 and USER F9 give error messages. *The Disc Drive* revealed an unlikely access method using POKE commands in BASIC to tweak the PCW's memory, sticking the byte value 229 into each of the addresses 66480, 64432, 64348 and 64040.

This was ingenious, but a bit cumbersome – loading up BASIC and all – and I thought it would be much more fun to pervert the PCW memory with a little Turbo Pascal routine. Turbo provides a built-in array called Mem which maps on to the program-area memory and can be read from or written to with an ordinary Pascal 'assign' statement:

```
Mem[66480]:= 229;
```

Oops! That gives an error warning, because the array index has to be a 16-bit

Integer variable, whose allowed range is -32,768 to 32,767. (Turbo Pascal version 4.0 introduced Word variables, with the often more useful range 0 to 65,535.) We have to tinker with the address numbers, subtracting 65,536 from each to get 'negative' values which look fanciful but do in fact work:

```
Program Limbo;
begin
  Mem[-1056]:= 229;
  Mem[-1104]:= 229;
  Mem[-1188]:= 229;
  Mem[-1496]:= 229;
end.
```

This compiles in Turbo Pascal to give LIMBO.COM, which when run in CP/M does indeed magically transport you to the secret F9 group – as conveyed by the fact that the prompt changes from A> to F9A>. Rod Shinkfield suggests deleting a copy of PIP.COM so it'll be available in the Limbo group, ready for a command like PIP DELETED.DOC[G0]=DELETED.DOC to copy a Limbo file back to group 0. A neat idea. You escape to A> normality with USER 0.

Unfortunately most people don't own Turbo Pascal; and even for those who do, the above still isn't the most elegant approach. This is because of the size of Turbo's run-time library, a collection of standard functions and procedures which means that even a program which literally does nothing at all will still take up 8.5k or so of disk space and spend much longer than necessary loading.

However, most (all?) owners of the old-style PCWs have an assembler program supplied as one of the CP/M utilities. So, obsessively, I looked up my mouldering grimoires of assembly language, and came up with a text file called LIMBO.ASM, containing the following:

```
POKE MACRO ADDR
LXI H,ADDR
MVI M,229
ENDM
POKE 64480
POKE 64432
POKE 64348
POKE 64040
JMP 0
```

The POKE macro loads the index register with the supplied address ADDR, and shoves 229 into that address. (I could equally well have written four pairs of LXI and MVI commands, but I'm showing off.) JMP 0 terminates the tiny program. I like the Turbo text editor for assembler programming; if you use LocoScript, the document needs to be converted to ASCII format.

With MAC.COM and HEXCOM.COM from the CP/M utilities disk, the command MAC LIMBO does the assembly, generating the intermediate file LIMBO.HEX – and HEXCOM LIMBO then produces LIMBO.COM. This tiny resulting program does exactly the same as the bulky Pascal one already described, but in 23 bytes rather than 8.5k.

Until, of course, driven by renewed obsessiveness, I found myself laboriously adding a BDOS 'print string' command to ensure that LIMBO.COM displays the all-important message 'Copyright (c) Ansible Information, 1999'. Yes, software is most definitely a disease.

PCW Today 13, March 2000

Me, *Ansible*, and Thog

Well, I have to admit it: my little software company Ansible Information is in its twilight years as far as the Amstrad PCW is concerned. Orders for AnsibleIndex, the famous LocoScript indexing software, are now rare and joyful things. Each time, it takes me longer and longer to remember how Ansible's home-made invoicing software works, and spiders have spun their webs over our last immemorial stack of 3" disks....

All this is something of a relief, since I've been ever so busy writing and occasionally editing – that other perennial subject of my old *PCW Plus* columns. Twenty years since I escaped the radioactive hell of being a weapons physicist at Aldermaston, I'm still a footloose freelance and struggling with some success not to end up in the gutter asking passers-by if they can spare a few Terry Pratchett paperbacks. Instead, I get paid for providing feedback on first drafts of Terry Pratchett's Discworld novels ... but that's another story.

Even the main hobby of what I laughingly call my spare time consists of writing and editing. This is the scurrilous science fiction newsletter *Ansible*, which is still the only non-North American SF fanzine ever to win the Hugo award, SF's equivalent of the Oscar – four times now. It's available free by e-mail and on the web, incidentally: see the links at www.ansible.co.uk. Strangely enough for a free publication, *Ansible* has opened doors leading me to sums of actual money.

Example: years ago the editor of Britain's longest-established SF fiction magazine *Interzone* decided he'd like a news and gossip page to provide variety amid all those worthy stories. I gave him an *Ansible* at a publishing party and he instantly commissioned a monthly column based on it. As I write, I've just sent in the 100th instalment.

Example: in 1995, by waving around all the vaguely prestigious Hugo awards resulting from my SF fanzine hobby (I also have fourteen as 'best fan writer', a peculiar category covering SF journalism, humour and gossip), I persuaded the newly founded magazine *SFX* to take on a regular Langford column. Nowadays *SFX* claims to be the world's best-selling SF magazine, which may even be true; and I'm working hard to persuade them that this is all because, at the time of writing, I've produced 70 columns without missing an issue.

Example: occasionally I scatter a few copies of *Ansible* in the SF sections of local Reading bookshops. This year someone passed it on to Waterstone's head office, and the editor of their planned on-line SF newsletter got in touch about reprinting some of the material for real money. We eventually settled on a version of the *Ansible* department called Thog's Masterclass ... which was also the subject of my column in the very last *PCW Plus*. The one the rotten sods at Future Publishing never sent me a complimentary copy of. Sniff.

What, asks the alert and keen-eyed reader, is Thog's Masterclass? Well, *Ansible* has long had a tradition of publishing awful or accidentally funny sentences from SF and fantasy – the more famous the author, the better. Meanwhile my pal John Grant had invented the huge and thick-witted barbarian Thog the Mighty for various of his fantasy novels. Somehow Thog's name attached itself to *Ansible*'s 'Ghastly Lines from Genre Fiction' department. The rest is history.

Thog's selections cover a wide range of fictional strangeness. Sometimes you suspect the author isn't entirely sure what he meant: 'They were both roughly the same age, in their very early fifties, though a hundred years earlier they would have appeared much younger.' (D.F. Jones, *Colossus*, 1966) Often fictional characters' eyes do unexpected things: 'They all felt Michael's adrenaline kick in and watched his eyes bounce off his legal pad ...' (Rock Brynner, *The Doomsday Report*, 1998) Certain authors have a flair for the utterly wrong simile: 'A silence descended like steel doors slamming down around the room.' (James P. Hogan, *Voyage from Yesteryear*, 1982)

Sheer overwriting offers treats like carnivorous weather: 'Rain came as a wet drizzle that clings to your face like a hungry leech fighting to hang on, only to slip down over the scars and dive into the abyss of excrement and refuse at your feet.' (Bradley Snow, *Andy*, 1990) Metaphors often fall awkwardly over one another: 'Only Lily could tell there was more to it, because whatever was haunting the back of his eyes made a trail of uneasy paw prints up her own spine.' (Charles de Lint, *Someplace to be Flying*, 1998) And it's so important to retain a sense of direction when piloting spacecraft: 'Captain Vandermeer, if you will please initiate a three-hundred-and-sixty-degree turn of the *Washington*, we'll begin the long journey home.' (Anne McCaffrey, *The Tower and the Hive*, 1999) There is more, all too much more.

Needless to say, I'm planning (with John Grant) to compile an entire book of Thog's Masterclass. The time is ripe, because electronic and print-on-demand publishing is making it easier for marginal or special-interest books to appear. One such publisher, Wildside Press, has already signed up my long out-of-print novel *Earthdoom* and its unpublished companion *Guts* (both, coincidentally, collaborations with John Grant), plus a huge collection of the 'Critical Mass' SF review columns I wrote in the 1980s. I wonder if they'd like a book of all my *PCW Plus* contributions?

PCW Today 14, October 2000

I Didn't Write That!

One nice thing about *PCW Today* is that our kindly editor doesn't muck around with what you write. Usually I dread reading my own bits in magazines or newspapers, for fear of finding what atrocities the subeditor has wrought.

In fact the worst disaster that ever befell a Langford column was accidental. This was in *8000 Plus* for January 1991, when I wrote up the horrors of the literary life as a joke PCW text adventure game whose player typed commands at the > prompt. Like this:

```
    You are in an indescribably sordid hallway. Shabbily carpeted
stairs lead up to your workroom.
    > HELP
    Kindly remember you are a freelance writer. That is: you're on
your own, sunshine.
    > GO NORTH
```

Stop kidding around. You have no idea which way is north.
[The rest of this piece can be found on page 105.]

At *8000 Plus* HQ it was an open secret that the magazine was designed on Macintosh computers. What I didn't know, and what the editors were too (let's be tactful) busy to notice, was that the Mac publishing software ignored lines starting with >. So only one side of that dialogue with the computer game was printed, lending the column a certain surrealism which made people ask what I'd been smoking that month.

Then came *The Guardian*, which hired me to write batches of quickie SF reviews with just 70 words allowed for each book. That wordcount included title, author and full publishing details, so *Eon* by Greg Bear would get a less cramped review than something like *Do Androids Dream of Electric Sheep* by Philip K. Dick, and I was grateful not to have to cover D.G. Compton's classic title *Hot Wireless Sets, Aspirin Tablets, The Sandpaper Sides of Used Matchboxes and Something That Might Have Been Castor Oil*.

Unfortunately, after I'd performed miracles of critical compression, the *Guardian* subeditors would then change everything around to fit the text into too small a space. Once they evened up their print columns by cutting a review's final sentence of praise for a book I'd liked, and sticking it on to my 70 words about one I'd hated. Don't believe everything you read in the newspapers.

Fortunately I've so far avoided the *Times Literary Supplement*. A fellow SF critic found that after commissioning a review, the poxy *TLS* would not only edit it savagely but would then pay only (say) 70% of the agreed fee, since they'd used only that percentage of a piece whose length had originally been specified with great precision. Shabby practice, indeed.

Then there's *SFX* magazine, for which I've been writing a column in every issue since it began in 1995. Even with that much experience and a page all to myself, I still can't judge the length of my contribution accurately enough to be safe from editorial cuts. This is because *SFX* is addicted to a trick of layout called the drop-quote, whereby the designer picks a phrase from your article that seems particularly brilliant or idiotic, and puts this in huge type inside a flashy box to make the page look more exciting.

Because you don't know how much space the drop-quote will take up, you can only guess at the needed word length. Every so often, a particularly oversized quote box forces cutting of less important stuff on the page. So my text gets trimmed, with the big knives homing in accurately on vital explanations and punchlines of jokes.

Most of the above cock-ups happen because of the mechanics of design and publishing, the never-ending tug-o'-war between text and presentation. Dorothy Sayers nailed this one back in 1933 when she set a detective novel in an advertising office and revealed 'that the great aim and object of the studio artist was to crowd the copy out of the advertisement and that, conversely, the copy-writer was a designing villain whose ambition was to cram the space with verbiage and leave no room for the sketch.' With electronic publishing, the same old battles are still fought on excitingly high-tech territory.

Well, you can be philosophical about all that. The worst annoyance comes when you meet an editor who wishes you'd written something different and tries to insert his ideas and opinions in place of yours. I had a nasty experience along these lines from *New Scientist* in the run-up to 2000....

It was a review feature based on futurology books. I made a point of digging out some less well-known differences between SF 'predictions' and reality, avoiding hackneyed cliches like the idea of food pills. The subeditor crossed out what I'd written and inserted 'food pills'. One of the books mentioned, and I quoted, an interesting bit about radical approaches to carbon dioxide sequestration – not planting trees in hope of soaking up greenhouse gases, but actually trapping and physically storing CO_2 emissions. The subeditor crossed that out and inserted 'planting trees'. I'd avoided quoting Arthur C. Clarke's far too familiar laws of futurology; the subeditor shoved them in, deftly deleting my own examples and conclusion to make room. Oh, and – being an SF fan myself – I didn't include the usual journalistic sneer at science fiction. The subeditor ... but you're ahead of me, aren't you? Never again.

Oddly enough, the next SF article I sold was to the seriously upmarket science journal *Nature*. Unlike dumbed-down *New Scientist*, they didn't change a word. I hope there's a moral there.

PCW Today 15, May 2001

Running Down

I keep wondering when I'll draw a final line under the extremely small software outfit Ansible Information, and retire to my other life of overwork as a writer. Certainly Ansible is beginning to flag, but – like the residue of a lingering head cold – it refuses to go entirely away.

2001 saw further decline when the immemorial Amstrad PCW 8512 lost the use of its B drive (no, changing the belt didn't help). A few months later, the A drive went too (ditto). This happened exactly as I was trying to move stuff from a pile of old 3" CF2 disks along a cable to the PC.

The result was a solemn pact between Ansible and SD Microsystems, whereby the great Steve Denson copied that particular batch of disks and in return will receive all future Ansible disk-transfer enquiries. Another fragment of the business empire successfully dismantled!

Further dismantling of a cruel and unusual nature was later carried out on the failed PCW plus a couple of dead ones that had accumulated in our cellar over the years. From the fragments, I managed to assemble a patchwork system with an evil, gaping hole where the B drive used to be. It's like owning Victor Frankenstein's home computer, born of unhallowed parts stolen from graveyards, and apt to turn at any moment on its hubristic creator. There's the terrible suspicion that Igor may have messed things up and brought me an abnormal processor chip.

Still, the undead PCW does actually work. For now.

Meanwhile, Microsoft continued its policy of causing as much irritation as possible to mere customers. Practically the only Ansible product that still sells is Ailink, the package that reads your 3½" PCW disks in a PC disk drive and then – this is the cunning part – converts the LocoScript documents to Rich Text Format files suitable for Word and other modern Windows word processors.

I have a suspicion that Bill Gates disapproves of this. At any rate, each version of the accursed Windows makes it harder for our software to read PCW CP/M disks on the PC. Back in the days of Windows 3.0 to 3.11 there was no problem. Windows 95 brought some difficulties: the CP/M copier needs to access disks in a special and nonstandard way which was now discouraged. With some personal loss of brain cells I programmed around this snag, but it got even worse in Windows 98.

Revelation! If you restart Windows 95 or Windows 98 in MS-DOS mode (an option on the Shut Down menu), all these difficulties with the CP/M Copier go away. Naturally Bill Gates got to hear of this, and came up with his cunning counterstroke. The new Windows ME couldn't be restarted in DOS mode.

Thanks to customer feedback, input from that nice Mr Denson, and a little lateral thinking, I came up with an all-new Ailink workaround that was far less drastic than restarting the computer, and seems to do the trick on Windows 95, 98 *and* ME.

It was at this point that the evil Gates, laughing satanically at the futile struggles of mere mortals, released Windows XP and set everything firmly back to square one. Maybe it would be a cunning plan to go out of business right now.

Meanwhile, although I'd vowed to have nothing to do with the non-PCW-compatible Amstrad PcW16, the inevitable finally happened in 2001. Someone with one of those little machines bought Ailink without heeding our subtly worded disclaimer ('If by any chance you need to convert PcW16 documents, PLEASE GO AWAY'). I was faced with the grim choice of bodging the PcW16 file format into Windows-readable shape, or offering a refund.

Naturally I chose the more complicated and time-consuming option. Just as Ansible's fabled software for the Amstrad PCW was written in Borland Turbo Pascal 3.0 for CP/M, its distant descendant Borland Delphi handles our current Windows 95+ programming. It was time to wield the mighty sledgehammer of Delphi upon the small and wizened nut of PcW16 document format.

Skipping over many tedious hours, I can report that PcW16 documents have a less utterly cryptic internal structure than LocoScript ones, although there are oddities that I haven't fathomed. The main obstacle to decoding them is what seems to be a bizarre attempt to save space by leaving out all the spaces (with mysterious exceptions). Instead, the ASCII coding of the first letter in each word (with the same mysterious exceptions) is increased by 127 as an indication that a space should be inserted just before. Very odd.

After various tweaks, the ad-hoc conversion to Windows Rich Text Format was producing recognizable prose and eliminating most of the formatting boilerplate. The bumf at the head of each PcW16 document varies in length: it was easier to delete by hand to the obvious start of the text than to automate this bit. A few inexplicable random letters still sprinkled the text, always at the beginnings of words, but these were obvious enough to the editorial eye.

So, another refund saved! It's not a saleable conversion program, though, because the results are scrappy and need hand-editing. All I need now is to discover that the full PcW16 document structure is publicly available on some website, and that other hands have already written a freeware conversion utility producing RTF results of unsurpassable perfection. This would provide suitable dramatic irony to conclude another episode in the decline and fall of Ansible Information.

PCW Today 16, February 2002 – the final column

Index

accounts 58
Adventure game, spoof 105, 181
advice column, spoof 66, 82, 132
advice, outdated 140
Airplane Game 125
Amstrad Portable (PPC) . . 56, 78, 87, 93
Amstrad support 15, 56, 136, 167
anonymous correspondents 155
Ansible Information 161, 183
Ansible Information, unlikely secrets of 79
Ansible, the newsletter 173
assembler 36
backups, failing to make 173
BASIC 17, 35, 167
book contracts 81
brand names 123
British SF Association 101
British Telecom 39, 114
bugs 25
cable link: wiring one's own 19
cat-vacuuming 136
celebrating 100 issues 152, 153
celebrating 50 issues 101
chain letters 124
Chandler, Raymond, spoofed 70
clip-art 137
computer addiction 49
computer languages 35
computer spares 56
computers as fantasy 49
computers in SF 12, 33
computers in SF: plots to avoid 41
contracts 81
copy-protection 59
copyediting 74
copyediting horror stories . 74, 165, 181
copyright 28
copyright law and titles 101
covering letter 54
CP/M 21, 85, 90
CP/M and MS-DOS 108
CP/M error messages 32
desktop publishing 137
disk drive repair 171
disk formats 134, 183
disk shortage 70
disk, boot-up 21

disk, start of day 21
disks 16, 85
dollar royalties in the UK 174
drinking games 72
DTP 137
Easter date calculation 150
edges and thresholds 94
Ediswan High-Voltage Healing Box . 158
editors on word-processed submissions 88
Encyclopaedia of Fantasy 159
Encyclopaedia of Science Fiction 138, 159
escape sequences 34
etiquette column 82
expository lumps 116
fan mail 26
fantasy trilogies, proliferation of 48
feminism 20
finance for freelances 57
Fizz-Buzz 72
FOG index 65
four-dimensional objects 164
freelancing and finance 57
freelancing horrors 105
future, a vision of 138
futurology 39
Gates, Bill, man of evil 183
GREASE 69
Herbert, A.P., spoofed 59
ideas: some sources 142
indexes and indexing 110
indexing software 161, 170
Internet 116
Internet, 1994 . . . . 146, 149
Joyce 11, 70
keyboard letters fading 15
keyboard scan oddities 12, 17
Langford column hiatus 146
Langford editorial form letter 128
laser printers 112
Laws of Robotics 33, 42
legal horrors 50
Leper's Squint 51
letter, covering 54
Life game 160
LocoScript 18, 50, 75, 170
LocoScript headers and footers 26
LocoScript templates 14

Lovecraftiana 163	search and replace 31, 38
magazine axed 170	self-employment 58
manuscript presentation . . . 11, 14, 128	serial communications 19
marketing software 119	sexism 20
Memorandum of Agreement 81	SF convention newsletter 144
metal things 122	SF Writers of America 44, 68
Milford SF Writers' Conference 100	shareware 27
Misleading Cases 59	slushpiles 37
Miss Magnetic Media 82	small claims court 103
MS-DOS and CP/M 108	Society of Authors 44, 97
Munchausen, Baron 165	software companies, how to annoy . . 24
Nebula Awards collection 68	software copy-protection 59
neologisms 52	software customers, how to annoy . . . 24
New Age movement 125	software marketing 63
newsletter production 144	software obsession 134, 178
obsolescence, technological 86	software pricing 22
occupational diseases 77	software reviews 52
opening sentences 147	software: edges and thresholds 94
openings and their impact 129	spreadsheets 175
padding and its joys 122	statistics 68
paperless office 13, 39, 73	style checkers 65
PCW monitor 13, 28	submission format 11, 14
phone calls, annoying 63	Sugar, Alan, and freebies 35
phrases, useful 140	Sugar, Alan, and tactical nukes 33
plagiarism and camels 29	Sugar, Alan, and unkind words 152
plot generators 143	Sugar, Alan, as serpent 41
postal persecution 155	Sugar, Alan, patter song 10
predictions for 1992 126	SuperWriter 18, 172
predictions in SF 130	technical support 90, 167
presentation 128	technofear 21, 90
Prestel 12, 34, 40, 115	technology, retro 157
print vs screen 50	Telecom Gold 40, 114
Protext 34	Thog's Masterclass 168, 180
Public Lending Right 61	typewriter magazine advice 66
publication and after 96	UFOs and lawsuits 91
publisher's reader 37	urban myth 45, 156, 167
publishing in crisis 117	Usenet 147
punctuation 75	Value Added Tax 47
pyramid selling 124	VAT 47
quotations 98	viruses 45
QWERTY keyboard 17	Windows 166, 183, 184
remaindering 96	word processors: devil's advocacy 30, 173
research techniques 163	WordStar 17, 22
reviewing books 164	Writers' and Artists' Yearbook 43
reviews, sycophantic 52	writers' workshops 99
royalties in dollars 174	writing about what you know 116
screen vs print 50	writing: how to avoid it 136

Printed in the United Kingdom by
Lightning Source UK Ltd., Milton Keynes
139362UK00001BA/6/P